MEDICAL
INTELLIGENCE
UNIT

Fas Signaling

Harald Wajant, Prof., Ph.D.

Department of Molecular Internal Medicine
Medical Clinic and Polyclinic II
University of Wuerzburg
Wuerzburg, Germany

LANDES BIOSCIENCE / EUREKAH.COM
GEORGETOWN, TEXAS
U.S.A.

SPRINGER SCIENCE+BUSINESS MEDIA
NEW YORK, NEW YORK
U.S.A.

FAS SIGNALING

Medical Intelligence Unit

Landes Bioscience
Springer Science+Business Media, Inc.

ISBN· 0-387-32172-1   Printed on acid-free paper.

Springer Science+Business Media, Inc., 233 Spring Street, New York, New York 10013, U.S.A.
http://www.springer.com

Please address all inquiries to the Publishers:
Landes Bioscience, 810 South Church Street, Georgetown, Texas 78626, U.S.A.
Phone: 512/ 863 7762; FAX: 512/ 863 0081
http://www.eurekah.com
http://www.landesbioscience.com

Printed in the United States of America.

9 8 7 6 5 4 3 2 1

Library of Congress Cataloging-in-Publication Data

Fas signaling / edited by Harald Wajant.
    p. ; cm. -- (Medical intelligence unit)
  Includes bibliographical references and index.
  ISBN 0-387-32172-1
 1. CD antigens. 2. Cellular signal transduction. I. Wajant, Harald. II. Title. III. Series: Medical intelligence unit (Unnumbered : 2003)
  [DNLM: 1. Antigens, CD95. 2. Adaptor Proteins, Signal Transducing. 3. Membrane Glycoproteins. QW 573 F248 2006]
  QR186.6.C42F37 2006
  571.7'4--dc22

                        2006004030

CONTENTS

EDITOR

Harald Wajant
Department of Molecular Internal Medicine
Medical Clinic and Polyclinic II
University of Wuerzburg
Wuerzburg, Germany
Email: harald.wajant@mail.uni-wuerzburg.de
Chapters 6, 10

CONTRIBUTORS

Nele Festjens
Molecular Signalling
  and Cell Death Unit
Department of Molecular Biology
Flanders Interuniversity Institute
  for Biotechnology
University of Gent
Gent, Belgium
Chapter 5

Consuelo Gajate
Centro de Investigacion del Cancer
Instituto de Biologia Molecular
  y Celular del Cancer
CSIC-Universidad de Salamanca,
  Campus Miguel de Unamuno
Salamanca, Spain
Chapter 2

Gregory J. Gores
Division of Gastroenterology
  and Hepatology
Mayo Medical School,
  Clinic and Foundation
Rochester, Minnesota, U.S.A.
Email: gores.gregory@mayo.edu
Chapter 8

Maria Eugenia Guicciardi
Division of Gastroenterology
  and Hepatology
Mayo Medical School,
  Clinic and Foundation
Rochester, Minnesota, U.S.A.
Chapter 8

Erich Gulbins
Department of Molecular Biology
University of Essen
Essen, Germany
Email: Erich.gulbins@uni-essen.de
Chapter 3

Georg Häcker
Institute for Medical Microbiology
Technische Universität München
Münich, Germany
Email: hacker@lrz.tu-muenchen.de
Chapter 9

Frank Henkler
Department of Molecular Medicine
Medical Clinic and Polyclinic II
University of Wuerzburg
Wuerzburg, Germany
Email:
  frank.henkler@mail.uni-wuerzburg.de
Chapter 10

Gabriele Hessler
Department of Molecular Biology
University of Essen
Essen, Germany
Chapter 3

Ottmar Janssen
Institute of Immunology
University of Kiel
Kiel, Germany
Email: ojanssen@email.uni-kiel.de
Chapter 7

Michael Kalai
Molecular Signalling
   and Cell Death Unit
Department of Molecular Biology
Flanders Interuniversity Institute
   for Biotechnology
University of Gent
Gent, Belgium
Chapter 5

Anja Krippner-Heidenreich
Institute of Cell Biology
   and Immunology
University of Stuttgart
Stuttgart, Germany
Email: anja.krippner@izi.uni-stuttgart.de
Chapter 1

Andreas Linkermann
Institute of Immunology
University of Kiel
Kiel, Germany
Chapter 7

Faustino Mollinedo
Centro de Investigacion del Cancer
Instituto de Biologia Molecular
   y Celular del Cancer
CSIC-Universidad de Salamanca
   Campus Miguel de Unamuno
Salamanca, Spain
Email: fmollin@usal.es
Chapter 2

Jing Qian
Institute of Immunology
University of Kiel
Kiel, Germany
Chapter 7

Xavier Saelens
Molecular Signalling
   and Cell Death Unit
Department of Molecular Biology
Flanders Interuniversity Institute
   for Biotechnology
University of Gent
Gent, Belgium
Chapter 5

Peter Scheurich
Institute of Cell Biology
   and Immunology
University of Stuttgart
Stuttgart, Germany
Email:
   peter.scheurich@po.uni-stuttgart.de
Chapter 1

Pascal Schneider
Institute of Biochemistry
University of Lausanne
Epalinges, Switzerland
Email: Pascal.Schneider@ib.unil.ch
Chapter 11

Volker Teichgräber
Department of Molecular Biology
University of Duisberg-Essen
Essen, Germany
Chapter 3

Margot Thome
Institute of Biochemistry
University of Lausanne
BIL Biomedical Research Center
Epalinges, Switzerland
Email: margot.thomemiazza@ib.unil.ch
Chapter 4

Tom Vanden Berghe
Department of Molecular Biology
Flanders Interuniversity Institute
   for Biotechnology
University of Gent
Gent, Belgium
Chapter 5

Peter Vandenabeele
Department of Molecular Biology
Flanders Interuniversity Institute
   for Biotechnology
University of Gent
Gent, Belgium
Email:
   Peter.Vandenabeele@dmb.rug.ac.be
Chapter 5

PREFACE

More than 15 years ago the isolation of monoclonal antibodies inducing programmed cell death (apoptosis) in lymphoblastic cells and fibroblasts by the groups of Peter Krammer (Heidelberg) and Minako Yonehara (Tokyo) opened the door to a new field of apoptosis research. The identification of Fas/anti-APO-1 (CD95), a member of the TNF receptor superfamily, as the antigen recognized by the death-inducing antibodies and the growing overall interest in apoptosis encouraged a huge number of researchers to take a closer look at this molecule and its functions. This work did not only lead to a detailed understanding of Fas-induced apoptosis and Fas biology but has also defined principle mechanisms which are of broader relevance. For example, the research on Fas revealed a new category of protein-protein interaction domains (the death domain fold), established the "induced proximity" model of caspase activation and also uncovered basic mechanisms of activation of TNF receptors. Moreover, Fas malfunction or deregulated Fas signaling have been implicated in a growing list of pathologies making Fas and its ligand (FasL) attractive targets for the development of new therapies. Recent studies revealed that Fas is more than just a death inducer. It has been found that Fas mediates liver regeneration, proliferation, neuronal differentiation and inflammation. Thus, the Fas field shows more and more diversification over the initial apoptosis-related focus.

We hope that this book will help students and scientists gain a general overview and first-hand information on Fas signaling and Fas biology from leading scientists in the field. In the first part of this book the signaling mechanisms of Fas are summarized whereas the second part of the book focuses on specialized aspects of Fas biology and discusses the medical and biotechnological relevance of the molecule.

Finally, I want to thank the authors of each chapter for their pleasant cooperation and the staff of Landes Bioscience, in particular Cynthia Conomos and Kristen Shumaker, for their great help during the whole editorial process.

Harald Wajant, Prof., Ph.D.

CHAPTER 1

FasL and Fas:

Typical Members of the TNF Ligand and Receptor Family

Anja Krippner-Heidenreich and Peter Scheurich

Abstract

The membrane receptor Fas is one of the central members of the TNF receptor super-family, representing the prototype of an apoptosis inducer. Its cognate ligand, FasL, is expressed as a type II transmembrane protein, but also exists as a soluble molecule. As typical for all members of the TNF receptor superfamily, Fas signaling is induced by multimerization. Both molecules share structural features with their respective family members. Hallmarks of Fas are two and a half copies of a cysteine rich domain in the extracellular part, required for ligand binding and self-association and the intracellular death domain, essential for apoptosis induction. Characteristics of the FasL are the stable formation of homotrimers, each protomer consisting of two β-sheets in a typical "jelly roll" arrangement and a proline-rich domain involved in intracellular transport and reverse signaling.

Introduction

The TNF Receptor and TNF Ligand Superfamilies

Fas (CD95/APO-1/TNFRSF6/APT1) is one of the best characterized members of the tumor necrosis factor receptor (TNFR) superfamily, currently comprising 29 receptors that are mirrored by only 19 ligands, representing the cognate TNF ligand superfamily. This already indicates that a single ligand might be capable to bind to more than one receptor and/or that there still exist orphan receptors (for review see refs. 1,2). Indeed, no ligand has so far been identified for DR6 (death receptor 6), RELT (receptor expressed in lympoid tissues) and TROY/TAJ. In addition, the ligands TNF, LT (lymphotoxin) alpha, FasL, VEGI (vascular endothelial cell-growth inhibitor), TRAIL (TNF-related apoptosis-inducing ligand), RANKL (receptor acitvator of nuclear factor-kappa B ligand), LIGHT, APRIL (a proliferation-inducing ligand) and BAFF (B-cell-activating factor) bind to more than one member of the TNFR superfamily. Some redundancy is present also in the Fas signaling system: FasL binds Fas as well as the soluble receptor DcR3 (decoy receptor 3), that might thus serve as a downregulator of Fas signaling by sequestration of its ligand (see below).

Additional crosstalk within the TNF receptor family occurs at the level of intracellular signal transducers. For example, the families of TRAF (TNF receptor associated factor) and

Fas Signaling, edited by Harald Wajant. ©2006 Landes Bioscience and Springer Science+Business Media.

death domain proteins contain members capable to interact with a number of TNF receptor family members.[3] As various TNF receptors can be found coexpressed e.g., in lymphoid cells, competition of these receptors for their intracellular adapter proteins is likely to occur. Moreover, proteins of the TRAF family also participate in signaling by other, functionally related receptors of the Toll-like IL-1 receptor superfamily.[4]

Most members of the TNFR family are expressed in the immune system and couple directly to signal pathways controlling activation, cell proliferation, survival and differentiation. In addition, these receptors orchestrate the (reversible) formation of multicellular structures like secondary lymphoid organs, hair follicles, teeth and bone.[2] They are crucially involved in protective functions during immune responses. Hallmarks of the activation of TNFR members are the induction of gene expression via members of the NF-κB (nuclear factor kappa B) transcription factor family, activation of MAP kinase family members like JNK (c-jun N-terminal kinase) and the stress activated kinase p38 as well as the induction of cell death.[2] Beside the typical "accidental" form of cell death called necrosis, leading to the release of cytoplasm into the intercellular space with a subsequent inflammatory response, a specialized, "silent" and altruistic form of cell death called apoptosis can be triggered by a subgroup of the TNFR members. Within this subgroup of apoptotic receptors Fas represents the prototype and is the best characterized member of the TNFR superfamily at the level of intracellular signal transduction. The phenomenon of apoptosis was described in the nineteenth century when scientists followed the morphological changes during the development of e.g., tadpoles and insects (for review see ref. 5). Since then this phenomenon was rediscovered several times until 1972 the term apoptosis (greek: apo, off, and ptosis, falling) was coined to describe a commonly observed type of programmed cell death.[5,6] However, the relevance of this process became evident only much later. Today it is well accepted that apoptosis is a process essential for life that is conserved from worm to mammal. Apoptosis plays an important role not only during embryogenesis, e.g., to shape organs, fingers and toes, but also in the development of the nervous system and the immune system. Furthermore, removal of unwanted cells in the adult individual is essential for tissue maintenance of multicellular organisms, especially to get rid of cells that are in excess (e.g., after a successful immune response) or potentially dangerous (e.g., upon development towards autonomous growth and malignancy; for review see ref. 7).

In this chapter we will give an overview on Fas and its ligand, FasL, focusing on their structure/function relationships.

Identification of Fas and Its Ligand

The Fas molecule was originally identified in 1989 by two independent groups. Trauth and coworkers described a monoclonal antibody, anti APO-1, raised against the lymphoblast cell line SKW6.4 that recognized a 52 kDa cell surface protein and strongly induced apoptosis in several cell lines.[8] Similar properties were described for the antibody anti-Fas by the group of Yonehara.[9] Intraperitoneal application of anti-Fas antibody in mice rapidly induced apoptotic destruction of the liver and death of the animals.[10] Molecular cloning of the antigen recognized by anti-Fas two years later revealed homologies to TNF receptor 1 and the additional members of the superfamily known at that time, CD40 and NGFR.[11] In 1993 the FasL was identified by expression cloning from a cytotoxic T cell line.[12] The sequence identified FasL as a type II transmembrane protein showing homology to TNF. A schematic representation of Fas and its ligand is given in Figure 1.

The Membrane Receptor Fas

The human Fas gene is located on chromosome 10q24.1 and consists of 9 exons spanning about 26 kb of DNA. The extracellular two and a half cysteine rich domains (see below) are encoded by 5 exons, i.e., the exon structure does not reflect the borders of these domains. This

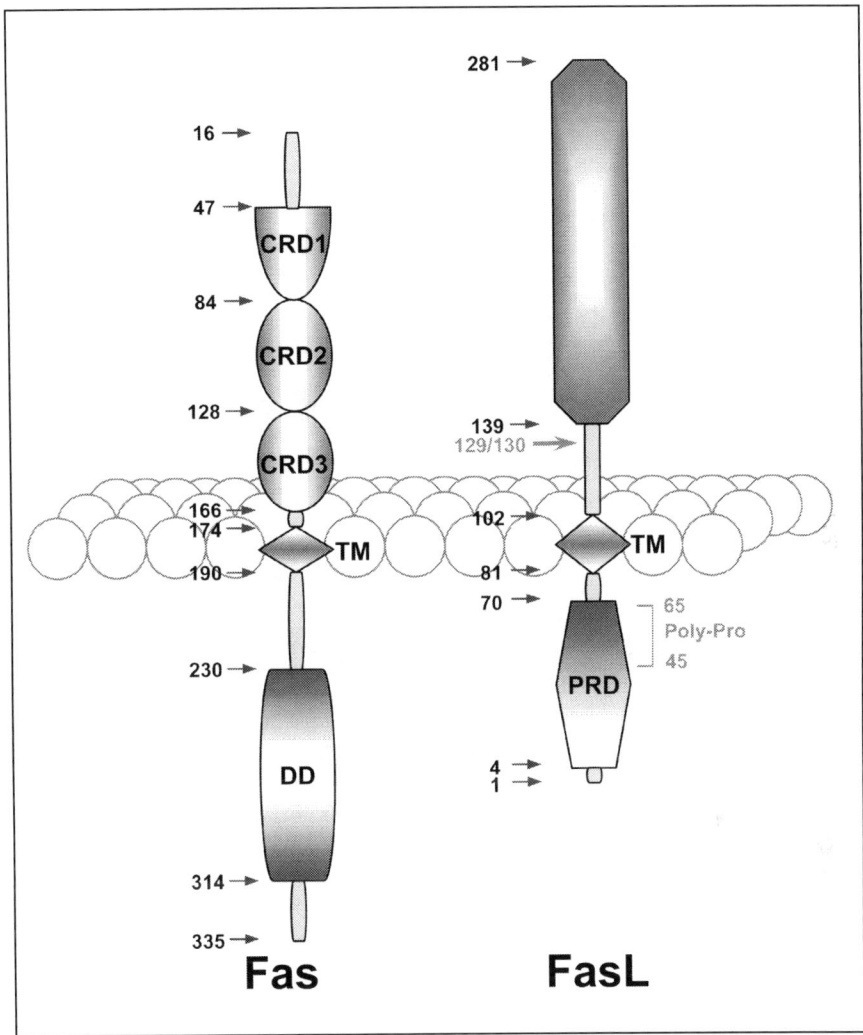

Figure 1. Schematic representation of Fas and FasL. Fas is a type I and FasL a type II transmebrane protein. The upper parts of both molecules represent the extracellular domains. CRD = cysteine rich domain; TM = transmembrane domain; DD = death domain; PRD = proline rich domain. Small arrows indicate amino acid numbers, the larger arrow the putative metalloproteinase cleavage site.

seems to be a common property of the whole TNF receptor family and has been described also in other members.[13] In contrast, the last exon, exon 9, fully comprises the C-terminal death domain of Fas. Analyses of the 5′ upstream sequence revealed no consensus for "TATA" and "CAAT" boxes at the appropriate positions rather than a GC-rich sequence.[13] This observation is consistent with multiple transcription initiation sites of Fas.

Fas is a type I glycoprotein with a single transmembrane domain, the typical structure of most members of the TNFR-superfamily. Cell surface expressed Fas consists of 319 aa with 157 aa comprising the extracellular and 145 aa the intracellular domain (Fig. 1). The extracellular portion of Fas is characterized by cysteine-rich domains (CRDs) (Fig. 1), a hallmark of all TNF receptor family members, containing at least one, but typically multiple of these structural units (for review see refs. 1,2). These domains consist of about 40 amino acids, and are often characterized by six cysteine residues forming intramolecular disulfide bonds. However, single CRDs can be further subdivided into smaller building blocks, called A, B and C modules, each CRD containing two of them at various combinations.[14] Using these definitions, the extracellular part of human Fas contains one incomplete CRD at the C-terminus consisting of only a single B module with a single disulfide bond. This is followed by two complete CRDs, with three and four disulfide bonds, respectively.[14] Accordingly, Fas might be described to possess two and a half CRDs, but there is some discrepancy concerning this point in the literature. A reason might be that the mouse homologue contains three full CRDs in the extracellular domain.[14] Functionally the intramolecular disulfide bonds stabilize the folding of the CRDs, in the case of Fas they are all required for receptor/ligand interaction.[15-17]

The complete CRDs 2 and 3 of Fas,[18] similar to CRD 2 and 3 of the respective four modules in TNFR1 and TNFR2,[19] form the sides of contact for ligand binding. As expected, these domains of Fas cannot be replaced with the respective domains of TNFR1 without loss of function. However, CRD1 of Fas as well as of TNFR1 is also required for ligand binding,[14,15,17] although displaying no ligand specificity. This is evident from experiments where CRD1 of TNFR1 could replace CRD1 of Fas and partially restore ligand binding that was abolished after CRD1 removal.[17] This suggests that CRD1 might be somehow involved in processes allowing ligand binding and/or receptor signaling, thus embodying a prerequisite for efficient ligand/receptor interaction. In fact, a recombinant protein comprising CRD1 of Fas has been described already some years ago to block Fas-mediated apoptosis. Later it could be shown that it forms homo-oligomers[20,21] and more recently the CRD1 of both Fas and TNFR1, as well as those of TNFR2, CD40 and DR4 (TRAIL-R1), were identified as domains allowing homophilic association of the respective receptors in the absence of ligand.[22,23] The responsible motif within CRD1, in Fas aa 1-49,[21] was therefore called the pre-ligand binding assembly domain (PLAD).[22,23] To date it is still unclear whether all members of the TNF receptor family contain a PLAD, as is its precise molecular mechanism in signal transduction. In a number of autoimmune Lymphoproliferative syndrome (ALPS) patients, carrying multiple mutations in the Fas gene, Siegel and colleagues found that the PLAD was preserved in all examples of dominant-negative mutations.[23] This implies that the mutant receptor causes dominant interference by physical interaction with wild type protein in a pre-associated receptor complex that normally permits Fas signaling.

In its intracellular part Fas contains an about 80 aa region, called the death domain (DD), that is characteristic for a subgroup of the TNFR superfamily, the death receptors (Fig. 1). As the name already implies these receptors are earmarked by their ability to induce apoptosis.[24,25] Besides Fas the group of death receptors comprises seven additional members (TNFR1, DR3 [death receptor 3], DR4 [TRAIL-R1 {TNF-related apoptosis-inducing ligand receptor 1}], DR5 [TRAIL-R2], DR6, p75-NGFR [p75-nerve growth factor receptor] and EDAR [ectodermal dysplasia receptor]). However, the current data from in vitro systems and knockout animals indicate a broad spectrum regarding the preferential cellular response of a particular death domain receptor.[26] Fas is considered as the prototypic apoptotic receptor, although it undoubtedly transmits nonapoptotic signals like the activation of NF-κB and JNK leading to gene expression (for review see Chapter 7). In contrast, the death domain receptors EDAR and DR6 have so far not been shown to represent strong apoptotic inducers, they rather activate NF-κB and MAP kinase members.[27,28]

Death domains are homophilic interaction domains. The three-dimensional structure of the Fas DD, as revealed by NMR spectroscopy, gave a first insight into the molecular structure of this signaling module.[29] It contains six antiparallel, amphiphatic alpha helices with an unusual topology. The surface of the Fas DD mainly consists of charged residues, suggesting that charge-charge interactions mediate complex formation between the DDs.[30] In fact, mere overexpression of many death domains mediates apoptosis and the Fas interacting signal transducer, FADD (Fas interacting DD; see below), binds Fas via homophilic DD interaction. The importance of the death domain for Fas signaling is underlined by the existence of naturally occurring mutant forms of mouse Fas termed lpr (lymphoproliferative[31]) and lprcg.[32] In lpr mice an early transposable element is inserted in intron 2 of the Fas gene, causing premature termination. Nevertheless full-length Fas mRNA can be detected as the lpr mutation is leaky. The lprgc allele encodes an isoleucine substituted for an asparagine at residue 225. This mutation within the DD abrogates the association with other DD proteins and therefore its biological activity too. Both mutations have a phenotype strongly dependent in strength on the background strain, but similar to gene-targeted Fas-null mice, resulting in the development of lymphoadenopathy, splenomegaly and increased levels of autoantibodies,[33] for review see ref. 34. By analogy to these mouse models also in patients several defects of the Fas pathway have been detected. Besides mutations in the Fas molecule and its ligand, also mutations concerning the signal transduction, e.g., in caspase 10, lead to the clinical symptoms of the childhood syndrome ALPS also known as Canale-Smith syndrome (CSS, for review see ref. 35).

The conserved fold of six α-helices in death domains is found also in other homophilic interaction domains, some of them being also involved in apoptotic signaling. These include the death effector domain (DED) and the caspase recruitment domain (CARD) as well as the pyrin domain.[36] The specificity of interaction within this superfamily of interaction domains is remarkable, no interaction between members of two subfamilies has been described so far.

Fas Expression

Many cell types, including epithelial cells, fibroblasts and hematopoietic cells, are known to express Fas. Notably, in the hematopoetic system Fas expression directly correlates with the maturation status of the cell. Whereas progenitor cells hardly express any Fas, cells were found to increase Fas expression gradually during differentiation.[37] Beside the full length mRNA of Fas in lymphocytes shorter mRNA species have been identified.[38,39] Six human splice variants have been described so far, all coding for proteins that lack the transmembrane domain of Fas but have retained the leader peptide and are thus secreted as soluble molecules. Most variants derive from a shift in the reading frame leading to premature stop codons. One of the variants contains all two and a half cysteine rich domains and should therefore retain full ligand binding capacity. Shorter variants, possibly lacking ligand binding capability, might still interfere with Fas signaling via PLAD – PLAD interaction with full length Fas (see above). In agreement with these considerations the splice variants have been described as downregulators of Fas mediated apoptosis, suggesting a physiological role as modulators of Fas signaling in general.[40] Unlike other members of the TNF receptor superfamily alternative splicing seems to be the major or only source of soluble Fas, besides proteolytic cleavage of the membrane expressed protein.

The Fas Ligand

The human FasL gene was mapped on chromosome 1q23 by in situ hybridization. The FasL gene consists of approximately 8.0 kb and is split into four exons.[41] FasL is predominantly expressed on activated T lymphocytes and natural killer cells but also at immune privileged sites (see Chapters 2, 10 and 12). Like most members of the TNF ligand family, FasL (CD178/ TNFSF6/APT1LG1) is a type II transmembrane protein and the bioactive form is a homotrimer (Fig. 1). Three potential N-glycosylation sites (N-X-S/T) exist in the extracellular domain that have been described to influence the expression level.[42] The 40 kDa protein shares about 15-35% amino acid identity with other members of the family with the highest homology in a 150 aa region in the extracellular domain of the ligands.[12] A typical structural feature of the TNF ligand family is the formation of a ß-sheet sandwich in the extracellular part consisting of ten amino acid strands in a "jelly roll" arrangement.[30,43] Five of these strands (aa 137-183 in FasL) form the hydrophobic interaction sites necessary for trimerization of the protomers. They are located in the inner part of the homotrimer and contain most of the conserved amino acids. The other five strands, mostly located in the C-terminus, form the outer β-sheet and are exposed to the extracellular fluid. They also form the contact sites for the respective receptors and carry less conserved amino acid residues.[12,42] Typically, the trimerized ligands possess a triangular conical structure with the N-terminus leading to the transmembrane domain at the base of this pyramidal structure. It is generally believed that also the membrane-integrated forms of the TNF family members form homotrimers as revealed from crystallographic studies of the soluble forms of TNF, LTα, CD40L und others, but so far not FasL.[43-46] Experiments with chemical crosslinkers, however, as well as sequence homologies between FasL and other TNF ligand family members strongly suggest that also FasL forms stable homotrimers as described above.

Another isoform of FasL lacks almost totally the extracellular domain and is therefore non functional in Fas binding. In addition it carries an altered aa sequence over a stretch of 10 aa in the residual extracellular domain (25 aa) (Swissprot P48023-2). The function of this FasL form has not been defined yet.

With the exception of the secreted cytokine LTα all TNF family members are (at least in part) produced as type II transmembrane proteins. The additionally observed soluble forms are produced by alternative splicing and/or proteolytic processing of the primary transmembrane products. Conversion of the membrane bound form of FasL to its soluble form is most likely catalyzed by cell-associated metalloproteinases, similar as described for TNF.[47-51] A putative protease cleavage site was identified C-terminal to K129, since deletion of amino acids 128-131 prevented the formation of soluble FasL.[47] As this putative cleavage sequence shares little homology to that of TNF or the CD40 ligand, the FasL processing enzymes are likely to be distinct from TACE (TNF alpha converting enzyme) known to process membrane bound TNF.[52,53] In addition, it has been shown that the metalloproteinases 3 (stromelysin-1) and 7 (matrilysin) are capable to process human FasL.[54,55] Different potential cleavage sites for matrilysin exist in human FasL,[56] but their usage in vivo is still undefined.

The intracellular domain of FasL comprises 80 amino acids and contains a proline-rich domain at aa 45 – 65 with several consensus SH3 binding motifs, that is responsible for sorting of FasL into secretory lysosomes as well as reverse signaling.[57,58] In analogy with other members of the TNF ligand family it is believed that FasL induces intracellular signals via its own cytoplasmic part upon receptor binding. This topic is covered in detail in chapter 8. In addition a putative casein kinase I motif is found in the intracellular part.

Analysis of the naturally occurring FasL mutant mice named gld (generalized lymphoproliferative disease) showed a point mutation in codon 273 resulting in an amino acid exchange of Phe to Leu close to the extracellular C-terminus of FasL. This mutant reveals no biological activity due to an attenuated receptor binding.[41] Similar to the lpr mice gld mice develop lymphoadenopathy, splenomegaly and show increased levels of autoantibodies (for review see ref. 34).

Membrane Bound and Soluble FasL

Similar to many other members of the TNF ligand superfamily, FasL can be cleaved by metalloproteinases in its extracellular domain to release the soluble trimeric ligand (see above).[12,59,60] Both forms are capable to bind to Fas, but only membrane bound FasL is efficient in induction of cytotoxicity.[61,62] As soluble FasL competes with the membrane-bound counterpart, however, it can act even as an antagonist preventing apoptosis induction by the membrane integrated form of the ligand.[47,48,61] On the other hand, soluble FasL binds effectively to fibronectin of the extracellular matrix, which results in retention of the molecule and an enhanced capacity to induce apoptosis.[63] In a similar way, also TNFR2, DR5 and CD40 are only fully activated by the membrane-bound form of their respective ligands.[64-67] However, the difference in bioactivity of the soluble vs. the membrane bound ligand form is not necessarily attributable to all receptors, as e.g., TNFR1 and DR4 are comparably well activated by both the soluble and the membrane-bound ligand.[64,68]

Besides its role in the blockage of apoptosis the soluble form of FasL has been shown to function as a strong chemoattractant and to enhance neutrophil and phagocyte migration to inflammatory sites.[69,70] Notably, this activity could be also mimicked by Fas specific agonistic antibodies, i.e., it is mediated by Fas, but was independent on a functional death domain of the receptor.[70] Moreover, ligand concentrations were effective, that were not able to significantly induce apoptosis.[69,70]

Receptor Ligand Complexes

As biologically active, soluble recombinant Fas ligand forms homotrimers under physiological conditions, it was assumed that the initial signal to induce intracellular signal transduction is the trimerization of three receptor molecules by the ligand.[30,71] In this complex, the elongated extracellular domains of Fas do not have direct contact, as revealed from the crystal structures of other ligand/receptor complexes, like that of the extracellular domain of TNFR1 bound by lymphotoxin alpha.[72] It is unclear, however, whether the cytoplasmic domains of the three receptors do interact directly, e.g., via their death domains. Indeed, in a docking model of Fas/FADD death domain complexes three different interaction sites have been proposed to exist on each death domain in two copies, i.e., a total number of 6 potential interaction sites per death domain. Tight packaging of the two domains of e.g., Fas and FADD would lead to the formation of regularly arranged, hexagonal death domain clusters.[73]

Originally it was assumed that unligated Fas molecules would exist as single molecules randomly distributed at the cell surface. However, as described above, the membrane distal cysteine rich domain of the extracellular part of Fas, the PLAD, mediates homophilic interaction. Crosslinking studies with bivalent chemical agents suggest that Fas, similar to other PLAD containing receptors, exists as preformed receptor trimers.[23] One might therefore speculate that PLAD-mediated trimerization of Fas, similar to the other PLAD positive receptors, pre-organizes these receptors in a way that (a) their capacity of ligand independent signaling is inhibited and/or (b) parallel binding of the trimerized ligand to all three receptors is allowed in a single step, rather than sequential binding of single receptor molecules in three steps. In fact, binding studies performed with iodinated ligands and PLAD positive receptors gave no evidence for distinct affinities that might be expected if complexes between one ligand and three receptor molecules are formed sequentially.[74,75]

The trimerization/reorganization of Fas molecules by its cognate ligand is accepted to represent an essential step for initiation of signal transduction, but is clearly not sufficient. Although soluble FasL is capable to trimerize its receptors, this form of the ligand has only poor bioactivity. In fact, development of strong intracellular signals has always been correlated with the development of large receptor/ligand clusters.[76] These findings resemble those also described in the TNF receptor system.[77] Formation of clusters can be extremely fast[78] and may be affected by the fluidity of the membrane and therefore by its lipid composition. Especially ceramide formation (see Chapter 4) has been described to enhance apoptotic signaling by Fas, and changes in the membrane composition favoring the formation of large Fas/FasL clusters may be one mechanism.[78,79] Internalization of receptor ligand complexes seems not to be mandatory for the induction of Fas mediated apoptosis,[76,80] but this might be different for the various TNF receptor family members.

A rationale for the need of large receptor clusters for efficient signaling has been given at two distinct levels: First, the group of Tschopp showed that hexamers of FasL possess superior apoptotic capacity when compared to normal, trimerized FasL. They argued that one ligated Fas trimer recruits only one molecule of caspase 8 from the cytoplasm (for details of Fas-induced signaling mechanisms, see chapters 5-7). Accordingly, at least two receptor trimer complexes, i.e., two molecules of caspase 8, are necessary for the initial proteolytic activation.[81] Second, it has been shown that the capacity of a ligand to strongly activate its cognate receptor strictly correlates with the stability of the ligand/receptor complexes. When these complexes dissociate fastly, as e.g., found for soluble TNF/TNFR2 binding, only a minor bioactivity is observed paralleled by the lack of formation of large receptor/ligand clusters in the membrane. Stable binding, in contrast, will lead to the formation of large clusters and strong intracellular signaling.[64,77,82] These observations favor the hypothesis that the kinetics of dissociation and association of receptor-ligand binding controls the formation of large clusters and thus the efficiency of signaling. The molecular mechanisms by which trimerized Fas bound to its ligand forms larger aggregates is still unknown. An attractive hypothesis is that PLAD/PLAD interactions serve to secondary group individual receptor/ligand complexes, but this has not been clarified so far.

Additional elements that regulate Fas complex formation/signaling are lipid rafts and intracellular structural components like the actin skeleton. In a recent and extensive study, the group of Peter has distinguished five stages of Fas signaling:[76] (1) The PLAD assembled, unligated Fas molecules, (2) Formation of submicroscopic Fas/FasL aggregates, (3) actin-dependent formation of the DISC (death inducing signaling complex), i.e., association of cytoplasmic components, (4) formation of microcopically visible clusters, this step is dependent on caspase activity, and, (5) internalization of the complexes. Whereas in these studies localization of Fas to lipid rafts was not required for efficient signaling,[76] in other cellular models a clear dependency was found.[78,79,83]

A number of reports exist where stress signals, like UV irradiation or treatment with chemotherapeutics result in Fas complex formation and subsequent Fas signaling, a mechanism that might be possible in the absence of FasL.[84-86]

Fas Signaling

Most aspects of Fas signaling are covered in other chapters of this book (see chapters 3-8) and therefore only a few, selected topics will be mentioned here. As described above, the stoichiometry of the Fas signaling complex appears to be multifaceted. At the molecular level it is widely accepted that the death domain of Fas transmits the apoptotic signal towards the caspase machinery via FADD. An alternative cytotoxic signaling pathway has been described by Tschopp and coworkers to occur in activated T lymphocytes. In these cells Fas stimulation induces a necrotic phenotype in the absence of significant caspase activation under critical involvement of the adaptor proteins FADD and RIP.[87]

The broad expression pattern of Fas suggests that many tissues are ready to activate the apoptotic program upon request from outside, i.e., via the extrinsic apoptotic pathway. However, the sole expression of Fas does not always confer sensitivity to Fas-mediated apoptosis, indicating additional control mechanisms at the level of intracellular signaling. Accordingly, Fas expression is a prerequisite for this type of apoptosis (and/or additional, non apoptotic signals) extrinsically triggered via FasL, but it is not sufficient.

Additional proteins have been described to directly bind the cytoplasmic domain of Fas. FAP-1, a protein tyrosine phosphatase, has been shown to bind the cytoplasmic C-terminal 15 amino acids of Fas and thus, block Fas signaling.[88,89] FAP-1 was found to colocalize with Fas at the Golgi complex that might prevent the translocation of Fas to the plasma membrane.[90] Conversely, the Fas-associated protein factor 1 (FAF-1) potentiates Fas-mediated apoptosis.

FAF-1 binds the cytoplasmic death domain of Fas as receptor mutants from lprcg mice are incapable to interact with this molecule.[91] Further, Bruton's tyrosine kinase, btk, has been described to interact with Fas, thereby blocking FADD binding and the induction of apoptosis.[92] An additional facet of Fas signaling and its regulation may occur by Fas association with additional membrane proteins. The human growth factor (HGF) receptor Met has been described to sequester Fas, thus blocking this molecule from FasL binding or autoaggregation.[93]

Beside soluble Fas isoforms (see above) an additional potential negative regulator for FasL signaling is decoy receptor 3 (DcR3/TR6),[94] also a member of the receptor superfamily, but not related to Fas. DcR3 was found to be amplified in about 50% of 35 primary lung and colon tumours investigated and DcR3 messenger RNA was expressed in malignant tissue.[95] These data strengthen its potential role as a functional decoy receptor. As DcR3 also binds two additional ligands, e.g., LIGHT and TL1A, there might exist some crosstalk between these signaling systems.[94,96]

References

1. Aggarwal BB. Signalling pathways of the TNF superfamily: a double-edged sword. Nat Rev Immunol 2003; 3:745-756.
2. Locksley RM, Killeen N, Lenardo MJ. The TNF and TNF receptor superfamilies: integrating mammalian biology. Cell 2001; 104:487-501.
3. Arch RH, Gedrich RW, Thompson CB. Tumor necrosis factor receptor-associated factors (TRAFs)-a family of adapter proteins that regulates life and death. Genes Dev 1998; 12:2821-2830.
4. Akira S, Takeda K, Kaisho T. Toll-like receptors: critical proteins linking innate and acquired immunity. Nat Immunol 2001; 2:675-680.
5. Lockshin RA, Osborne B, Zakeri Z. Cell death in the third millennium. Cell Death Differ 2000; 7:2-7.
6. Kerr JF, Wyllie AH, Currie AR. Apoptosis: a basic biological phenomenon with wide-ranging implications in tissue kinetics. Br J Cancer 1972; 26:239-257.
7. Nagata S. Apoptosis by death factor. Cell 1997; 88:355-365.
8. Trauth BC, Klas C, Peters AM et al. Monoclonal antibody-mediated tumor regression by induction of apoptosis. Science 1989; 245:301-305.
9. Yonehara S, Ishii A, Yonehara M. A cell-killing monoclonal antibody (anti-Fas) to a cell surface antigen co-downregulated with the receptor of tumor necrosis factor. J Exp Med 1989; 169:1747-1756.
10. Ogasawara J, Watanabe-Fukunaga R, Adachi M et al. Lethal effect of the anti-Fas antibody in mice. Nature 1993; 364:806-809.
11. Itoh N, Yonehara S, Ishii A et al. The polypeptide encoded by the cDNA for human cell surface antigen Fas can mediate apoptosis. Cell 1991; 66:233-243.
12. Suda T, Takahashi T, Golstein P et al. Molecular cloning and expression of the Fas ligand, a novel member of the tumor necrosis factor family. Cell 1993; 75:1169-1178.
13. Cheng J, Liu C, Koopman WJ et al. Characterization of human Fas gene. Exon/intron organization and promoter region. J Immunol 1995; 154:1239-1245.
14. Naismith JH, Sprang SR. Modularity in the TNF-receptor family. Trends Biochem Sci 1998; 23:74-79.
15. Marsters SA, Frutkin AD, Simpson NJ et al. Identification of cysteine-rich domains of the type 1 tumor necrosis factor receptor involved in ligand binding. J Biol Chem 1992; 267:5747-5750.
16. Yan H, Chao MV. Disruption of cysteine-rich repeats of the p75 nerve growth factor receptor leads to loss of ligand binding. J Biol Chem 1991; 266:12099-12104.
17. Orlinick JR, Vaishnaw A, Elkon KB et al. Requirement of cysteine-rich repeats of the Fas receptor for binding by the Fas ligand. J Biol Chem 1997; 272:28889-28894.
18. Starling GC, Bajorath J, Emswiler J et al. Identification of amino acid residues important for ligand binding to Fas. J Exp Med 1997; 185:1487-1492.

19. Banner DW, D'Arcy A, Janes W et al. Crystal structure of the soluble human 55 kd TNF receptor-human TNF beta complex: implications for TNF receptor activation. Cell 1993; 73:431-445.

20. Papoff G, Cascino I, Eramo A et al. An N-terminal domain shared by Fas/Apo-1 (CD95) soluble variants prevents cell death in vitro. J Immunol 1996; 156:4622-4630.

21. Papoff G, Hausler P, Eramo A et al. Identification and characterization of a ligand-independent oligomerization domain in the extracellular region of the CD95 death receptor. J Biol Chem 1999; 274:38241-38250.

22. Chan FK, Chun HJ, Zheng L et al. A domain in TNF receptors that mediates ligand-independent receptor assembly and signaling. Science 2000; 288:2351-2354.

23. Siegel RM, Frederiksen JK, Zacharias DA et al. Fas preassociation required for apoptosis signaling and dominant inhibition by pathogenic mutations. Science 2000; 288:2354-2357.

24. Itoh N, Nagata S. A novel protein domain required for apoptosis. Mutational analysis of human Fas antigen. J Biol Chem 1993; 268:10932-10937.

25. Tartaglia LA, Ayres TM, Wong GH et al. A novel domain within the 55 kd TNF receptor signals cell death. Cell 1993; 74:845-853.

26. Wajant H. Death receptors. Essays Biochem 2003; 39:53-71.

27. Pan G, Bauer JH, Haridas V et al. Identification and functional characterization of DR6, a novel death domain-containing. TNF receptor FEBS Lett 1998; 431:351-356.

28. Koppinen P, Pispa J, Laurikkala J et al. Signaling and subcellular localization of the TNF receptor Edar. Exp Cell Res 2001; 269:180-192.

29. Huang B, Eberstadt M, Olejniczak ET et al. NMR structure and mutagenesis of the Fas (APO-1/ CD95) death domain. Nature 1996; 384:638-641.

30. Fesik SW. Insights into programmed cell death through structural biology. Cell 2000; 103:273-282.

31. Watanabe-Fukunaga R, Brannan CI, Copeland NG et al. Lymphoproliferation disorder in mice explained by defects in Fas antigen that mediates apoptosis. Nature 1992; 356:314-317.

32. Matsuzawa A, Moriyama T, Kaneko T et al. A new allele of the lpr locus, lprcg, that complements the gld gene in induction of lymphadenopathy in the mouse. J Exp Med 1990; 171:519-531.

33. Adachi M, Suematsu S, Kondo T et al. Targeted mutation in the Fas gene causes hyperplasia in peripheral lymphoid organs and liver. Nat Genet 1995; 11:294-300.

34. Nagata S, Suda T. Fas and Fas ligand: lpr and gld mutations. Immunol Today 1995; 16:39-43.

35. Rieux-Laucat F, Le Deist F, Fischer A. Autoimmune lymphoproliferative syndromes: genetic defects of apoptosis pathways. Cell Death Differ 2003; 10:124-133.

36. Lahm A, Paradisi A, Green DR et al. Death fold domain interaction in apoptosis. Cell Death Differ 2003; 10:10-12.

37. Stahnke K, Hecker S, Kohne E et al. CD95 (APO-1/FAS)-mediated apoptosis in cytokine-activated hematopoietic cells. Exp Hematol 1998; 26:844-850.

38. Cheng J, Zhou T, Liu C et al. Protection from Fas-mediated apoptosis by a soluble form of the Fas molecule. Science 1994; 263:1759-1762.

39. Cascino I, Fiucci G, Papoff G et al. Three functional soluble forms of the human apoptosis-inducing Fas molecule are produced by alternative splicing. J Immunol 1995; 154:2706-2713.

40. Ruberti G, Cascino I, Papoff G et al. Fas splicing variants and their effect on apoptosis. Adv Exp Med Biol 1996; 406:125-134.

41. Takahashi T, Tanaka M, Brannan CI et al. Generalized lymphoproliferative disease in mice, caused by a point mutation in the Fas ligand. Cell 1994; 76:969-976.

42. Orlinick JR, Elkon KB, Chao MV. Separate domains of the human fas ligand dictate self-association and receptor binding. J Biol Chem 1997; 272:32221-32229.

43. Eck MJ, Sprang SR. The structure of tumor necrosis factor-alpha at 2.6 A resolution. Implications for receptor binding. J Biol Chem 1989; 264:17595-17605.

44. Eck MJ, Beutler B, Kuo G et al. Crystallization of trimeric recombinant human tumor necrosis factor (cachectin). J Biol Chem 1988; 263:12816-12819.

45. Jones EY, Stuart DI, Walker NP. Structure of tumour necrosis factor. Nature 1989; 338:225-228.

46. Eck MJ, Ultsch M, Rinderknecht E et al. The structure of human lymphotoxin (tumor necrosis factor-beta) at 1.9- A resolution. J Biol Chem 1992; 267:2119-2122.

47. Tanaka M, Itai T, Adachi M et al. Downregulation of Fas ligand by shedding. Nat Med 1998; 4:31-36.

48. Suda T, Hashimoto H, Tanaka M et al. Membrane Fas ligand kills human peripheral blood T lymphocytes, and soluble Fas ligand blocks the killing. J Exp Med 1997; 186:2045-2050.
49. Adachi M, Suematsu S, Suda T et al. Enhanced and accelerated lymphoproliferation in Fas-null mice. Proc Natl Acad Sci USA 1996; 93:2131-2136.
50. Kayagaki N, Kawasaki A, Ebata T et al. Metalloproteinase-mediated release of human Fas ligand. J Exp Med 1995; 182:1777-1783.
51. Mariani SM, Matiba B, Baumler C et al. Regulation of cell surface APO-1/Fas (CD95) ligand expression by metalloproteases. Eur J Immunol 1995; 25:2303-2307.
52. Black RA, Rauch CT, Kozlosky CJ et al. A metalloproteinase disintegrin that releases tumour-necrosis factor- alpha from cells. Nature 1997; 385:729-733.
53. Moss ML, Jin SL, Milla ME et al. Cloning of a disintegrin metalloproteinase that processes precursor tumour-necrosis factor-alpha. Nature 1997; 385:733-736.
54. Matsuno H, Yudoh K, Watanabe Y et al. Stromelysin-1 (MMP-3) in synovial fluid of patients with rheumatoid arthritis has potential to cleave membrane bound Fas ligand. J Rheumatol 2001; 28:22-28.
55. Powell WC, Fingleton B, Wilson CL et al. The metalloproteinase matrilysin proteolytically generates active soluble Fas ligand and potentiates epithelial cell apoptosis. Curr Biol 1999; 9:1441-1447.
56. Vargo-Gogola T, Crawford HC, Fingleton B et al. Identification of novel matrix metalloproteinase-7 (matrilysin) cleavage sites in murine and human Fas ligand. Arch Biochem Biophys 2002; 408:155-161.
57. Wenzel J, Sanzenbacher R, Ghadimi M et al. Multiple interactions of the cytosolic polyproline region of the CD95 ligand: hints for the reverse signal transduction capacity of a death factor. FEBS Lett 2001; 509:255-262.
58. Blott EJ, Bossi G, Clark R et al. Fas ligand is targeted to secretory lysosomes via a proline-rich domain in its cytoplasmic tail. J Cell Sci 2001; 114:2405-2416.
59. Suda T, Nagata S. Purification and characterization of the Fas-ligand that induces apoptosis. J Exp Med 1994; 179:873-879.
60. Tanaka M, Suda T, Takahashi T et al. Expression of the functional soluble form of human fas ligand in activated lymphocytes. EMBO J 1995; 14:1129-1135.
61. Schneider P, Holler N, Bodmer JL et al. Conversion of membrane-bound Fas(CD95) ligand to its soluble form is associated with downregulation of its proapoptotic activity and loss of liver toxicity. J Exp Med 1998; 187:1205-1213.
62. Shudo K, Kinoshita K, Imamura R et al. The membrane-bound but not the soluble form of human Fas ligand is responsible for its inflammatory activity. Eur J Immunol 2001; 31:2504-2511.
63. Aoki K, Kurooka M, Chen JJ et al. Extracellular matrix interacts with soluble CD95L: retention and enhancement of cytotoxicity. Nat Immunol 2001; 2:333-337.
64. Grell M, Douni E, Wajant H et al. The transmembrane form of tumor necrosis factor is the prime activating ligand of the 80 kDa tumor necrosis factor receptor. Cell 1995; 83:793-802.
65. LeBlanc HN, Ashkenazi A. Apo2L/TRAIL and its death and decoy receptors. Cell Death Differ 2003; 10:66-75.
66. Muhlenbeck F, Schneider P, Bodmer JL et al. TRAIL-R1 and TRAIL-R2 have distinct cross-linking requirements for initiation of apoptosis and are non-redundant in JNK activation. J Biol Chem 2000.
67. Haswell LE, Glennie MJAl, Shamkhani A. Analysis of the oligomeric requirement for signaling by CD40 using soluble multimeric forms of its ligand, CD154. Eur J Immunol 2001; 31:3094-3100.
68. Wajant H, Moosmayer D, Wuest T et al. Differential activation of TRAIL-R1 and -2 by soluble and membrane TRAIL allows selective surface antigen-directed activation of TRAIL-R2 by a soluble TRAIL derivative. Oncogene 2001; 20:4101-4106.
69. Ottonello L, Tortolina G, Amelotti M et al. Soluble Fas ligand is chemotactic for human neutrophilic polymorphonuclear leukocytes. J Immunol 1999; 162:3601-3606.
70. Seino K, Iwabuchi K, Kayagaki N et al. Chemotactic activity of soluble Fas ligand against phagocytes. J Immunol 1998; 161:4484-4488.
71. Smith CA, Farrah T, Goodwin RG. The TNF receptor superfamily of cellular and viral proteins: activation, costimulation, and death. Cell 1994; 76:959-962.
72. D'Arcy A, Banner DW, Janes W et al. Crystallization and preliminary crystallographic analysis of a TNF-beta- 55 kDa TNF receptor complex. J Mol Biol 1993; 229:555-557.

73. Weber CH, Vincenz C. A docking model of key components of the DISC complex: death domain superfamily interactions redefined. FEBS Lett 2001; 492:171-176.
74. Kull FC Jr., Jacobs S, Cuatrecasas P. Cellular receptor for 125I-labeled tumor necrosis factor: specific binding, affinity labeling, and relationship to sensitivity. Proc Natl Acad Sci USA 1985; 82:5756-5760.
75. Scheurich P, Ucer U, Kronke M et al. Quantification and characterization of high-affinity membrane receptors for tumor necrosis factor on human leukemic cell lines. Int J Cancer 1986; 38:127-133.
76. Algeciras-Schimnich A, Shen L, Barnhart BC et al. Molecular ordering of the initial signaling events of CD95. Mol Cell Biol 2002; 22:207-220.
77. Krippner-Heidenreich A, Tubing F, Bryde S et al. Control of receptor-induced signaling complex formation by the kinetics of ligand/receptor interaction. J Biol Chem 2002; 277:44155-44163.
78. Cremesti A, Paris F, Grassme H et al. Ceramide enables fas to cap and kill. J Biol Chem 2001; 276:23954-23961.
79. Grassme H, Cremesti A, Kolesnick R et al. Ceramide-mediated clustering is required for CD95-DISC formation. Oncogene 2003; 22:5457-5470.
80. Schutze S, Machleidt T, Adam D et al. Inhibition of receptor internalization by monodansylcadaverine selectively blocks p55 tumor necrosis factor receptor death domain signaling. J Biol Chem 1999; 274:10203-10212.
81. Holler N, Tardivel A, Kovacsovics-Bankowski M et al. Two adjacent trimeric Fas ligands are required for Fas signaling and formation of a death-inducing signaling complex. Mol Cell Biol 2003; 23:1428-1440.
82. Grell M, Wajant H, Zimmermann G et al. The type 1 receptor (CD120a) is the high-affinity receptor for soluble tumor necrosis factor. Proc Natl Acad Sci USA 1998; 95:570-575.
83. Hueber AO, Bernard AM, Herincs Z et al. An essential role for membrane rafts in the initiation of Fas/CD95- triggered cell death in mouse thymocytes. EMBO Rep 2002; 3:190-196.
84. Rehemtulla A, Hamilton CA, Chinnaiyan AM et al. Ultraviolet radiation-induced apoptosis is mediated by activation of CD- 95 (Fas/APO-1). J Biol Chem 1997; 272:25783-25786.
85. Friesen C, Herr I, Krammer PH et al. Involvement of the CD95 (APO-1/FAS) receptor/ligand system in drug- induced apoptosis in leukemia cells. Nat Med 1996; 2:574-577.
86. Petak I, Houghton JA. Shared pathways: death receptors and cytotoxic drugs in cancer therapy. Pathol Oncol Res 2001; 7:95-106.
87. Holler N, Zaru R, Micheau O et al. Fas triggers an alternative, caspase-8-independent cell death pathway using the kinase RIP as effector molecule. Nat Immunol 2000; 1:489-495.
88. Sato T, Irie S, Kitada S et al. FAP-1: a protein tyrosine phosphatase that associates with Fas. Science 1995; 268:411-415.
89. Yanagisawa J, Takahashi M, Kanki H et al. The molecular interaction of Fas and FAP-1. A tripeptide blocker of human Fas interaction with FAP-1 promotes Fas-induced apoptosis. J Biol Chem 1997; 272:8539-8545.
90. Ungefroren H, Kruse ML, Trauzold A et al. FAP-1 in pancreatic cancer cells: functional and mechanistic studies on its inhibitory role in CD95-mediated apoptosis. J Cell Sci 2001; 114:2735-2746.
91. Chu K, Niu X, Williams LT. A Fas-associated protein factor, FAF1, potentiates Fas-mediated apoptosis. Proc Natl Acad Sci USA 1995; 92:11894-11898.
92. Vassilev A, Ozer Z, Navara C et al. Bruton's tyrosine kinase as an inhibitor of the Fas/CD95 death-inducing signaling complex. J Biol Chem 1999; 274:1646-1656.
93. Wang X, DeFrances MC, Dai Y et al. A mechanism of cell survival: sequestration of Fas by the HGF receptor. Met Mol Cell 2002; 9:411-421.
94. Yu KY, Kwon B, Ni J et al. A newly identified member of tumor necrosis factor receptor superfamily (TR6) suppresses LIGHT-mediated apoptosis. J Biol Chem 1999; 274:13733-13736.
95. Pitti RM, Marsters SA, Lawrence DA et al. Genomic amplification of a decoy receptor for Fas ligand in lung and colon cancer. Nature 1998; 396:699-703.
96. Migone TS, Zhang J, Luo X et al. TL1A is a TNF-like ligand for DR3 and TR6/DcR3 and functions as a T cell costimulator. Immunity 2002; 16:479-492.

CHAPTER 2

FasL-Independent Activation of Fas

Faustino Mollinedo and Consuelo Gajate

Abstract

Fas death receptor (also named CD95 or APO-1) is physiologically activated through binding to its cognate ligand, FasL. Fas/FasL interaction induces oligomerization and aggregation of Fas receptor, leading eventually to apoptosis after protein-protein interactions with adaptor and effector proteins. However, recent evidences demonstrate that either oligomerization of the receptor in trimers, as well as Fas aggregation in large clusters do not require its interaction with FasL. Activation of Fas through its translocation into membrane rafts, forming Fas caps, can be rendered independently of FasL. This FasL-independent cocapping of Fas in membrane rafts generates high local concentrations of Fas, providing scaffolds for coupling adaptor and effector proteins involved in Fas signaling. Thus, Fas receptor can be modulated either extracellularly, via FasL, or intracellularly independently of its ligand. Unraveling the molecular mechanism involved in FasL-independent activation of Fas will raise putative novel therapeutic interventions, especially in disorders where apoptosis is deficient such as cancer and autoimmune diseases, avoiding in this way the deleterious side effects that preclude the use of systemic activation of the Fas receptor by its ligand.

Abbreviations

The abbreviations used are: DED, death effector domain; DEF, death effector filament; DISC, death-inducing signaling complex; ET-18-OCH$_3$, 1-O-octadecyl-2-O-methyl-rac-glycero-3-phosphocholine (edelfosine); FADD, Fas-associated death domain-containing protein; FasL, Fas ligand; JNK, c-Jun NH$_2$-terminal kinase; lpr, lymphoproliferation; MKK7, mitogen-activated protein kinase kinase 7; TNF, tumor necrosis factor.

Introduction

In May 1989 Dr. Shin Yonehara et al[1,2] reported an IgM monoclonal antibody that could kill several human cell lines, and termed Fas (FS7-associated cell surface antigen) the cell surface protein recognized by the antibody. In July of the same year, Dr. Peter H. Krammer and his associates reported a mouse monoclonal antibody, named anti-APO-1 antibody which promoted apoptosis in human leukemic cells and activated lymphocytes.[3] In 1991 Dr. Shigekazu Nagata and his associates succeeded in cloning the membrane protein recognized by the killing antibody, the Fas antigen,[4] that turned out to be identical to the APO-1 protein identified later in Krammer's group.[5] Then Nagata's group cloned the corresponding physiological ligand of the Fas death receptor, named Fas ligand (FasL).[6-8] These findings identified the Fas/FasL system as the major regulator of apoptosis at the cell membrane in mammalian cells through a

Fas Signaling, edited by Harald Wajant. ©2006 Landes Bioscience and Springer Science+Business Media.

receptor/ligand interaction. Subsequent studies led to a detailed characterization of the initial events triggered after binding of Fas to its cognate ligand. However, the notion that Fas requires interaction with its ligand to trigger an apoptotic response has been recently challenged. Current evidence indicates that Fas can be activated in a FasL-independent manner, suggesting that Fas can also receive orders from within the cell. Thus, cells could dictate its own demise through activation of Fas death receptor without the external influence of its ligand. Unraveling the signaling pathways and molecules involved in the intracellular activation of Fas, independently of FasL, can provide the basis for novel therapeutic strategies and for the development of new compounds able to modulate apoptosis.

Fas Death Receptor Functions through Oligomerization and a Cascade of Protein-Protein Interactions

The Fas death receptor (also called CD95 or APO-1), a major member of the tumor necrosis factor (TNF)/nerve growth factor receptor family[4,9] transmits apoptosis signals initiated by the interaction with its membrane-bound or soluble natural ligand FasL or by agonistic anti-Fas antibodies.[1,3] FasL belongs to the TNF family[6] and can be found as a 40-kDa membrane-bound or a 26-kDa soluble cytokine.[10]

Mature Fas (Fig. 1) is a 45-kDa type I transmembrane receptor of 319 amino acids with a single transmembrane domain of 17 amino acids (from Leu-158 to Val-174), an N-terminal cysteine-rich extracellular domain (18 cysteine residues in 157 amino acids) and a C-terminal cytoplasmic domain of 145 amino acids that is relatively abundant in charged amino acids (24 basic and 19 acidic amino acids). The cytoplasmic portion of Fas contains a domain of about 85 amino acids termed "death domain", which is homologous to other death receptors and plays a crucial role in transmitting the death signal from the cell's surface to intracellular pathways.[4,9,11] Unlike the intracellular regions of other transmembrane receptors involved in signal transduction, the death domain does not possess enzymatic activity, but mediates signaling through protein-protein interactions. The death domain has the propensity to self-associate and form large aggregates in solution. The tertiary structure of the Fas death domain, revealed by NMR spectroscopy,[12] consists of six antiparallel, amphipathic α helices.[12,13] Helices α1 and α2 are centrally located and flanked on each side by α3/α4 and α5/α6. This leads to an unusual topology in which the loops connecting α1/α2 and α4/α5 cross over each other.[12,13] The presence of a high number of charged amino acids in the surface of the death domain is probably responsible for mediating the interactions between death domains. Stimulation of Fas by FasL results in receptor aggregation,[14,15] previously assembled in trimers,[16,17] and recruitment of the adaptor molecule Fas-associated death domain-containing protein (FADD)[18] through interaction between its own death domain and the clustered receptor death domains. FADD also contains a "death effector domain" (DED) that binds to an analogous domain repeated in tandem within the zymogen form of caspase-8.[19] Upon recruitment by FADD, procaspase-8 oligomerization drives its activation through self-cleavage, activating downstream effector caspases and leading to apoptosis.[20] Thus, activation of Fas results in receptor aggregation and formation of the so-called "death-inducing signaling complex" (DISC),[21] containing trimerized Fas, FADD and procaspase-8 (Fig. 2).

Mice carrying the lymphoproliferation (lpr) point mutation which converts Ile-225 to Asn-225 in the cytoplasmic region of the mouse Fas antigen, are characterized by a deficient Fas antigen that leads to a lymphoproliferation syndrome showing lymphadenopathy and a systemic lupus erythematosus-like autoimmune disease.[22] The corresponding mutation in human Fas (V238N) leads to inhibition of apoptosis,[11] together with a dramatic inhibition in Fas death domain self-association and binding to FADD,[12] suggesting that this point mutation alters the protein structure of the death domain. These data suggest that the intracellular portion of the Fas molecule is critical for death receptor oligomerization required for apoptotic

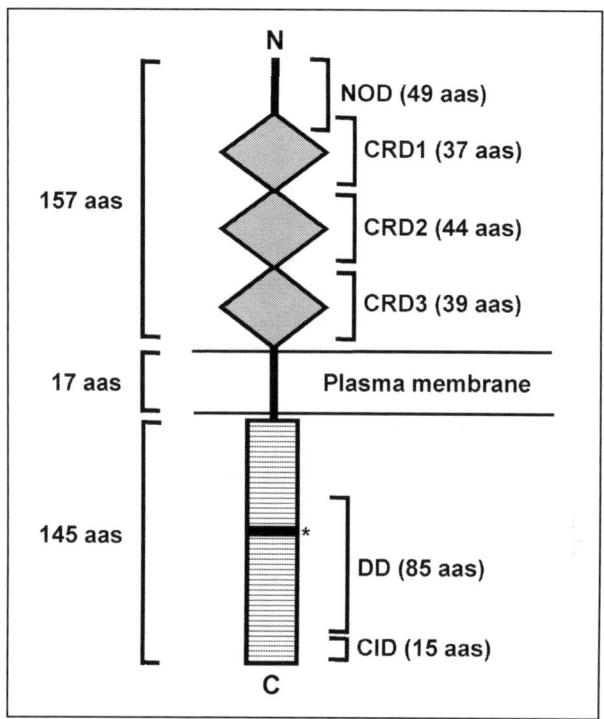

Figure 1. Schematic diagram of the human Fas death receptor. Mature human Fas protein consists of 319 amino acids (aas) with an N-terminal extracellular domain of 157 aas, a short transmembrane region (17 aas) and a C-terminal cytoplasmic domain of 145 aas. Relevant domains for Fas oligomerization and apoptotic activity are shown. An N-terminal extracellular oligomerization domain (NOD) of 49 aas (Arg-1 to Pro-49) responsible for FasL-independent oligomerization of the receptor. Three cysteine-rich domains (CRD1 -Gln31 to Val-67, CRD2 -Pro-68 to Cys-111-, and CRD3 -Arg-112 to Lys150-) containing 4, 6 and 8 Cys residues in each domain respectively. A cytoplasmic death domain (DD) of 85 aas (Ser-214 to Ile -298) is crucial for oligomerization and apoptotic signaling. The last 15 amino acids (Asp-305 to Val-319) of the Fas amino acid sequence represent a C-terminal inhibitory domain (CID). The corresponding *lpr* point mutation in human Fas (Val238N) is also shown (*). Domains and membrane are not to scale. Amino acid numbering is according to Itoh et al[4] for the mature human Fas sequence.

activity. Current evidence indicates that the molecular ordering of the initial events in physiological Fas-mediated signaling include four successive steps.[23] (a) FasL-induced formation of Fas microaggregates at the cell surface, (b) recruitment of FADD to form a DISC in an actin filament-dependent manner, (c) formation of large Fas surface clusters positively regulated by DISC-generated caspase-8, (d) actin filament-dependent internalization of activated Fas through an endosomal pathway.

Fas Oligomerization without Interaction with Its Ligand FasL

An early view of the molecular events leading to Fas activation considered that, upon binding to homotrimers of FasL, the Fas receptor homo-oligomerized through the intracellular death domains resulting in its trimerization. Due to the propensity of the intracellular Fas death domains to associate with one another, FasL ligation led to the clustering of Fas death domains and Fas aggregation in trimers promoting the recruitment of FADD and procaspase-8 that together with the receptor formed the DISC and triggered apoptosis.[21]

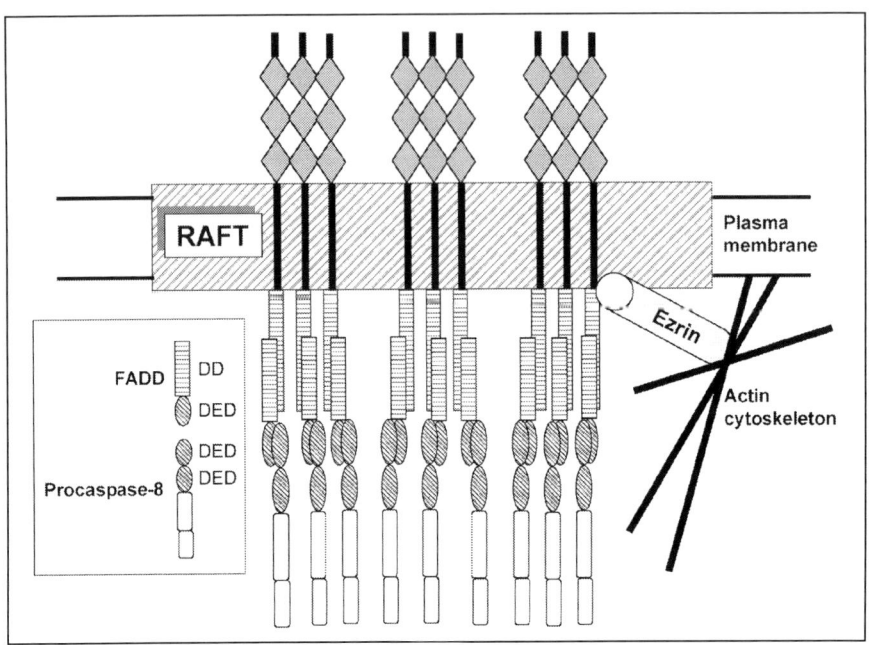

Figure 2. Schematic representation of a tentative model for activation of Fas receptor through its aggregation in membrane rafts. Fas molecules are brought together and concentrate in membrane rafts facilitating the formation of DISCs, following protein-protein interactions between Fas-FADD through their respective death domains (DD), and FADD-procaspase-8 through their respective death effector domains (DED). Actin cytoskeleton through ezrin could be involved in the clustering of Fas in rafts.

However, experimental evidences accumulated in the last five years have modified our view of how Fas-signaling activation takes place. Recent evidence suggests that Fas receptor is in trimer status before FasL binding.[16,17] Preassociated Fas complexes were found in living cells by means of fluorescence resonance energy transfer between variants of green fluorescent protein.[17] A FasL- and death domain-independent oligomerization domain in the extracellular region of the Fas receptor, mapping to the NH_2-terminal 49 amino acids, mediates homo- and hetero-oligomerization of the death receptor[16] (Fig. 1). Thus, Fas in unstimulated cells seems to be constitutively oligomerized and ligand binding is suggested to induce conformational changes of oligomerized subunits relative to each other, bringing together intracellular domains that are separated in the basal state. However, apoptosis can be triggered in the absence of FasL by overexpressing the Fas cytoplasmic domain or a Fas receptor lacking the NH_2-terminal 42 amino acids,[16] suggesting that the extracellular oligomerization domain of Fas is not required to initiate signaling and that self-association of the death domain is necessary and sufficient to induce cell death and occurs in the absence of an intact extracellular oligomerization domain. Thus, two major oligomerization domains seem to be present in the Fas receptor (Fig. 1), one mapping to the extracellular region of the receptor, likely related to the regulation of the nonsignaling state, and another one, involved in apoptotic signaling, mapping to the intracytoplasmic region, the death domain.[16]

The intracellular death domains of death receptors show a high tendency to self-associate[24,25] and when overexpressed by gene transfer in eukaryotic cells trigger signaling for cytotoxicity.[25] These findings indicate that the Fas receptor plays an active role in its own clustering and suggest the existence of cellular mechanisms that restrict its self-association, thus preventing constitutive signaling. Altogether, these data show that Fas oligomerization can be rendered in the absence of FasL.

Fas-Mediated Apoptosis Does Not Require the Participation of FasL

It has been proposed that doxorubicin-induced apoptosis in human T-leukemic cells is mediated by FasL expression with subsequent autocrine and/or paracrine induction of cell death through binding of FasL to the membrane Fas receptor.[26] This led to postulate in 1996 that Fas/FasL interactions could account for chemotherapy-associated apoptosis.[26] In addition to doxorubicin, additional anticancer drugs, such as methotrexate or bleomycin, were reported to promote induction of FasL expression and upregulation of membrane FasL, leading to autocrine or paracrine Fas/FasL-dependent apoptosis.[26-29] Activation-induced cell death (AICD) in T cells as well as cell death promoted by different antitumor drugs have been reported to be mediated by Fas/FasL interaction.[30,31] Cell lines resistant to Fas were reported to be insensitive to anticancer drug-induced apoptosis, and drug-induced cell death was prevented by Fas neutralizing antibodies.[26] However, the above notion involving Fas/FasL interactions in chemotherapy-induced apoptosis became rapidly controversial as several research groups did not obtain the same results, and thereby did not reach the same conclusions as above, using even similar experimental conditions that included same cell types and chemotherapeutic drugs.[32-38] Furthermore, a number of reports showed that blockade of Fas/FasL interactions did not prevent apoptosis induced by doxorubicin and additional cytotoxic drugs.[32-38] Although many cytotoxic drugs have been reported to act independently of the Fas system,[34-36,39,40] we and others detected FasL-independent activation of Fas in the mechanism of action of a number of antitumor drugs, including the antitumor ether lipid 1-*O*-octadecyl-2-*O*-methyl-*rac*-glycero-3-phosphocholine (ET-18-OCH$_3$, edelfosine), cisplatin (CDDP), etoposide (VP16), and vinblastine.[38,41] Cells deficient in Fas were resistant to the proapoptotic action of ET-18-OCH$_3$, but became sensitive to the antitumor ether lipid when transfected with Fas.[38] The presence or absence of Fas cell surface expression in cancer target cells correlated with their sensitivity or resistance, respectively, to the proapoptotic activity of ET-18-OCH$_3$.[38] Down-regulation of FADD by transient transfection of an antisense FADD construct inhibited tumor cell sensitivity to cisplatin, etoposide or vinblastine, whereas overexpression of FADD sensitized tumor cells to drug-induced cell death.[41] Transfection of cells with FADD dominant negative decreased apoptosis induced both by cisplatin or antitumor ether lipids,[41,42] and transient transfection with either MC159 or E8, two viral proteins that inhibit apoptosis at the level of FADD and procaspase-8, respectively,[43] protected cells from cisplatin-induced cytotoxicity.[41] Nevertheless, incubation with blocking anti-Fas antibodies (such as ZB4 and SM1/23 antibodies) or with the soluble Fas-IgG chimera fusion protein to block the interaction of Fas with FasL failed to inhibit drug-induced apoptosis, and drug-mediated induction of FasL expression was not always detected in distinct tumor cells.[38,41] Thus, these data suggest that, at least, some anticancer drugs induce cell death through a Fas/FADD pathway in a FasL-independent manner, and can partially explain the previous contradictory results on the involvement of Fas signaling in the action of anticancer drugs.

On the other hand, the unique mechanism of action of the antitumor ether lipid ET-18-OCH$_3$ demonstrated the intracellular activation of Fas independently of FasL.[38,44,45] This antitumor ether lipid mediates apoptosis in cancer cells through Fas activation once the drug is inside the cell, and normal cells are spared because they are unable to incorporate significant amounts of the drug. Fas-expressing cells that do not take up ET-18-OCH$_3$ from the culture medium are unaffected by the ether lipid when this latter is added exogenously, but they undergo rapid apoptosis following microinjection of ET-18-OCH$_3$ into the target cell.[38]

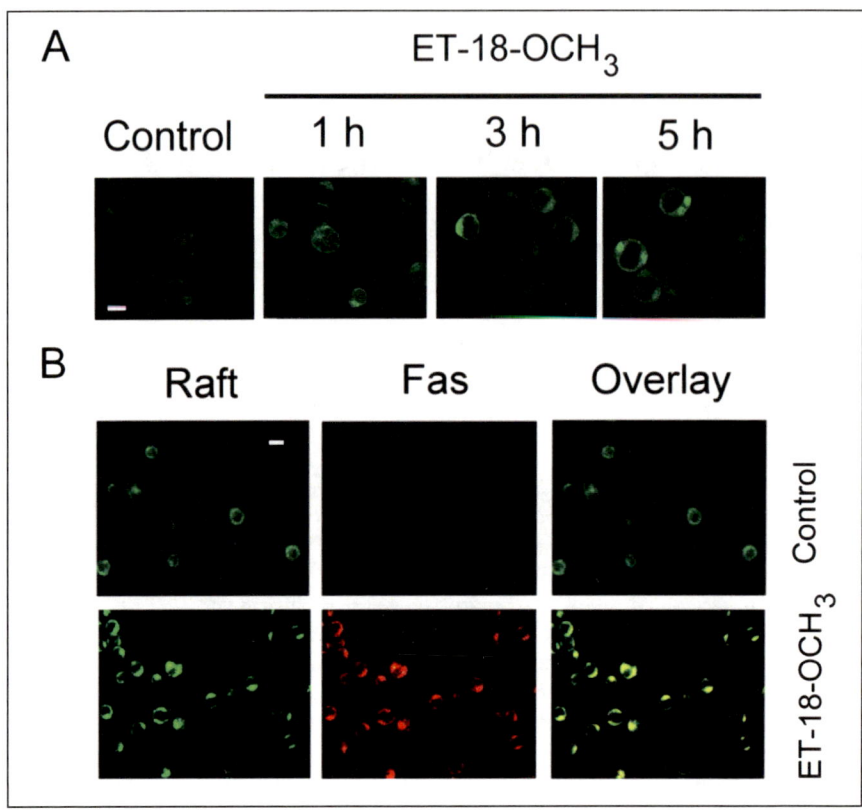

Figure 3. Cocapping of membrane rafts and Fas in human leukemic Jurkat cells treated with the antitumor ether lipid ET-18-OCH$_3$. A) Time-course of the effect of ET-18-OCH$_3$ on aggregation of membrane rafts. T-leukemic Jurkat cells were either untreated (Control) or treated with 5 µg/mL ET-18-OCH$_3$ for the times indicated. Cells were then stained with fluorescein isothiocyanate-labeled cholera toxin B subunit (FITC-CTx) used as a raft marker, and analyzed by confocal microscopy. *Bar*, 7 µm. B) Colocalization of membrane rafts (Raft) and Fas in ET-18-OCH$_3$-treated Jurkat cells. Cells were either untreated (Control) or treated with ET-18-OCH$_3$ for 3 hours, and processed for confocal microscopy using FITC-CTx (green fluorescence for lipid rafts) and anti-Fas monoclonal antibody, followed by CY3-conjugated anti-mouse antibody (red fluorescence for Fas). Areas of colocalization between membrane rafts and Fas in the overlay panels are yellow. *Bar*, 10 µm. (From C. Gajate and F. Mollinedo. Blood 2001; 98:3860-3863. © American Society of Hematology, used with permission, ref. 46.)

FasL-Independent Translocation and Capping of Fas into Membrane Rafts

While investigating the mechanism of action of the antitumor ether lipid ET-18-OCH$_3$ we found that this drug induced apoptosis in leukemic cells through translocation and capping of Fas into membrane rafts (Fig. 3), independently of FasL.[46] This was assessed by both confocal microscopy and isolation of membrane rafts through sucrose gradient centrifugation.[46] Raft disruption inhibited both ET-18-OCH$_3$-induced Fas capping and apoptosis.[46] Thus, these data involved for the first time membrane rafts in Fas-mediated apoptosis and cancer chemotherapy. Subsequent studies also found that Fas was translocated into lipid rafts following activation with FasL, as well as recruitment of FADD and procaspase-8 to the rafts forming the so-called DISC.[47,48] However, the translocation and capping of Fas into membrane rafts following ET-18-OCH$_3$ treatment was independent of FasL.[38,46] In a recent study, resveratrol, a polyphenol found mainly in grape skin with antitumor chemopreventive properties, has also been found to redistribute Fas in rafts independently of FasL.[49]

Membrane rafts are membrane microdomains consisting of dynamic assemblies of cholesterol and sphingolipids.[50-54] The presence of saturated hydrocarbon chains in sphingolipids allows for cholesterol to be tightly intercalated, leading to the presence of distinct liquid-ordered phases, membrane rafts, dispersed in the liquid-disordered matrix, and thereby more fluid, lipid bilayer.[55] One key property of membrane rafts is that they can include or exclude proteins to varying degrees. Membrane rafts may serve as foci for recruitment and concentration of signaling molecules at the plasma membrane, and thus they have been implicated in signal transduction from cell surface receptors.[51]

We hypothesize that accumulation of Fas into aggregates of stabilized membrane lipid domains from a highly dispersed distribution may represent a general mode of regulating Fas activation. Thus, membrane rafts could serve, in addition to generate high local concentration of Fas, as platforms for coupling adaptor and effector proteins required for Fas signaling (Fig. 2). This is of particular importance in Fas-mediated signal transduction as most of the initial signaling events depend largely on protein-protein interactions. Furthermore, this could facilitate and amplify signaling processes by local assembly of various cross-interacting signaling molecules (Gajate et al, in preparation).

Interestingly the lipid molecule ceramide has been involved in the clustering of Fas into ceramide-rich rafts.[56,57] However, ceramide acts as a mediator of the clustering process not as an initiator of the process, amplifying the primary Fas signaling events. Thus, C$_{16}$-ceramide did not trigger Fas clustering in the absence of a stimulatory anti-Fas antibody or FasL.[57] It is suggested[56-60] that Fas-FasL complexes enter initially into small membrane rafts and induce a weak formation of the DISC leading to caspase 8 activation. This rather weak caspase 8 activation then would generate ceramide through sphingomyelinase translocation and activation to the small lipid rafts. Due to the high amount of sphingomyelin present in rafts (as much as 70% of all cellular sphingomyelin may be found in rafts[61]), the generated ceramide could induce coalescence of elementary rafts[62] leading to the formation of big patches containing Fas-FasL complexes that would further lead to enhanced DISC formation, thereby potentiating Fas signaling. Thus, sphingomyelinase and ceramide serve to amplify the signaling of Fas at the membrane level after the initial Fas-FasL interaction.

However, the current view supports the notion that Fas/FasL interaction, although can enhance cell death,[63,64] is not essential in drug-induced apoptosis, and a growing number of agents and experimental conditions can induce Fas activation without the participation of FasL (Table 1). As recent evidence indicates that Fas clustering can occur without the participation of FasL, it could be suggested that either the different treatments exerting Fas clustering promote sphingomyelinase-dependent ceramide generation or induce physical changes in the plasma membrane similar to those elicited by ceramide inducing coalescence of rafts leading to large raft platforms and subsequent capping. Thus, activation of Fas by FasL or in a ligand-independent fashion leads to a visible aggregation of Fas under the microscope as compared with untreated cells, supporting that the Fas aggregation observed[38,41,65-69] is in a macroscopic level.

FasL-Independent Activation of Fas by Unrelated Agents

In the last few years a number of reports have evidenced that Fas can be activated under conditions that preclude interaction with FasL. Table 1 lists the agents reported so far to cause FasL-independent activation and clustering of Fas. These data suggest a common mechanism whereby divergent stimuli can activate membrane-associated events that target the Fas apoptotic

Table 1. FasL-independent activation of Fas by different agents

	Experimental Evidence		
Inducer	FasL-Independent Activation of Fas	Fas Clustering/ Capping	Cocapping of Fas and Rafts
Camptothecin	ref. 103		
CDDP	ref. 41	ref. 41	
Curcumin	ref. 71	ref. 71	
ET-18-OCH3	ref. 38, 46	ref. 38, 46	ref. 46
Glutamine deprivation-mediated cell shrinkage	ref. 104	ref. 104	
HCV core protein	ref. 105	ref. 105	
JNK activation (via MKK7)	ref. 69	ref. 69	
Resveratrol	ref. 49	ref. 49	ref. 49
TGF-β1	ref. 106		
TK/GCV	ref. 70	ref. 70	
Ultraviolet light	ref. 65, 67, 107	ref. 65, 67, 107	
Vanadate	ref. 72	ref. 72	
Vinblastine	ref. 41	ref. 41	
VP16	ref. 41	ref. 41	

Fas clustering or aggregation was visualized by immunofluorescence confocal microscopy or assessed by immunoprecipitating Fas using limiting antibody concentrations.[65] Co-capping of Fas and rafts was visualized by immunofluorescence confocal microscopy.[46] CDDP, cisplatin. HCV, hepatitis C virus. TK/GCV, herpes simplex thymidine kinase/ganciclovir. VP16, etoposide. TGF-β1, transforming growth factor-β1.

pathway in a manner that precludes its natural ligand FasL. This FasL-independent activation of Fas is blocked by dominant negative-FADD, antisense FADD, caspase-8 inhibitors, or by MC159 and E8, that inhibit the FADD/caspase-8 pathway, involving FADD and caspase-8 in this process.[41,49,69-72] On these grounds it is clear that this Fas clustering induces apoptosis in a FADD-mediated way, thus suggesting that Fas clustering promotes FADD recruitment and DISC formation, independently of FasL (Fig. 2).

Formation of platforms where a large amount of signaling molecules are brought together increases DISC formation and therefore potentiates Fas signaling. Because activation of caspase-8 is induced by proximity,[73] its concentration in lipid rafts will favor caspase-8 activation, triggering the downstream apoptotic signaling. It can be envisaged that the intrinsic enzymatic activity of caspase-8, upon approximation of additional procaspase-8 molecules mediated by the adapter FADD molecules, attains a sufficient concentration to activate the apoptosis pathway. Using chimeras of caspase-8 with either CD8 or Tac, Martin and coworkers[74] found that oligomerization at the cell membrane powerfully induces caspase-8 autoactivation and apoptosis. On these grounds, it can be envisaged that these oligomerization processes would be facilitated enormously in the large Fas aggregates formed during stimulation, leading to activation of caspase-8 and the ensuing generation of downstream apoptotic signals.

Thus, Fas clustering could be an efficient way to elicit apoptosis through recruitment of the DED-containing proteins FADD and caspase-8 into Fas caps (Fig. 2). In addition, it has also been demonstrated that FADD and caspase-8 coalesce into what appear to be perinuclear "death effector filaments" (DEFs) inducing receptor-independent apoptotic signals and apoptosis.[75,76] Overexpression of either FADD or caspase-8 induces apoptosis through the formation of unique filament structures that contain the death effector domains of these proteins,[75,76] being named accordingly "death effector filaments" (DEFs). Thus, formation of death effector filaments leads to intracellular assemblies of apoptosis-signaling complexes that can initiate or amplify apoptotic stimuli independently of receptors at the plasma membrane. Cycloheximide has been shown to induce cell death in human leukemic Jurkat and CEM C7 T-cell lines in a FADD-dependent and receptor-independent manner through DEF formation.[68] Also, a number of antitumor drugs, including microtubule-disrupting agents, may induce apoptosis via caspase-8 activation independently of the Fas/FasL system.[77]

As stated above the initial events in Fas signaling are largely dependent on the local concentration of the three major components of DISC, either Fas, FADD and caspase-8, and in fact, oligomerization of each one are able to mount an apoptotic response. Thus, formation of Fas caps leads to the recruitment of these molecules in a limited space, increasing the probability of interactions among them, and thereby promoting a strong apoptotic response.

How Are Fas Clusters Formed from Internal Signals?

As shown in Table 1, FasL-independent activation is mediated by Fas clustering, and recent evidence shows cocapping of Fas in membrane rafts.[46,49] How are these Fas clusters generated? As FasL is not strictly required, signals from inside the cell must regulate this process. The formation of Fas clusters as well as the recruitment of Fas into membrane rafts in a FasL-independent manner could involve intracellular processes, changes in the physicochemical properties of cell membranes or both.

Vanadate-elicited Fas aggregation and Fas-FADD association, as well as caspase-8 activation, were dependent on c-Jun NH_2-terminal kinase (JNK) activation,[72] as assessed by the use of the selective JNK inhibitor D-JNKI1.[78] These results highlight a major role for JNK in the signaling mechanisms leading to FasL-independent Fas activation. In fact, selective JNK activation by overexpressing the mitogen-activated protein kinase kinase 7 (MKK7) induced cell death mediated by FADD and Fas activation, independently of FasL.[69] Persistent JNK activation led to clustering of Fas.[69] Other inducers of FasL-independent Fas capping lead to a rapid and persistent activation of JNK, such as the antitumor ether lipid ET-18-OCH$_3$[79] and vanadate.[72] These data suggest that persistent JNK activation could be at least one of the signaling events leading to Fas clustering. In this regard, ceramide, which also favors Fas aggregation, induces apoptosis through sustained JNK activation.[80] However, the molecular events between JNK activation and Fas clustering remain to be elucidated. Persistent JNK activation is linked to cell death induced by different agents and stress conditions,[79,81-84] and its critical role in apoptosis is also supported by the lack of cell death on hippocampal neurons in JNK3-deficient and in JNK1/JNK2 double knockout mice.[85,86] In addition JNK-mediated apoptosis could involve mitochondria as JNK can translocate into mitochondria promoting phosphorylation and inactivation of anti-apoptotic Bcl-2 and Bcl-x$_L$.[87,88] Furthermore, UV-induced activation of the mitochondrial-mediated death pathway is abrogated in the absence of JNK, further supporting mitochondria as the target of JNK.[89]

Another putative mechanism involved in Fas clustering could involve cytoskeleton, a dynamic intracellular structure that due to its continuous assembly/disassembly could be perfectly equip to translocate proteins and transmit signals.[90] The interactions between plasma membrane and cytoskeleton play an essential role in various cellular functions,[91-93] and a link between raft-mediated signaling and the interaction of actin cytoskeleton with raft membrane domains has been suggested.[94] Ezrin, a major protein of the so-called ERM proteins (ezrin, radixin, moesin) linking the actin cytoskeleton to the plasma membrane,[95,96] interacts with Fas and mediates Fas cell membrane polarization (Fig. 2) during Fas-induced apoptosis in human T lymphocytes.[97] In addition, disialoganglioside GD3 redistributes in membrane-associated domains colocalizing with ezrin in Fas-triggered apoptosis, and GD3 is present in ezrin immunoprecipitates.[98] On the other hand, changes in cytoskeleton and microtubule disruption activate JNK,[90,99,100] but whether persistent JNK activation could induce reorganization of the microtubule cytoskeleton that promotes Fas clustering remains to be determined.

Concluding Remarks

Recent evidences indicate that Fas signaling is mediated by the formation of large Fas aggregates. Under physiological conditions FasL triggers Fas aggregation in caps. Nevertheless, this capping can be also generated by nonphysiological agents without the participation of FasL (Table 1), raising the possibility for new therapeutic interventions. This is of interest due to the toxic side effects derived from the use of FasL or agonistic anti-Fas antibodies in vivo leading to a fatal hepatic damage.[101,102] When an agonistic anti-Fas antibody or recombinant FasL was injected into mice to activate the Fas system in vivo, the mice were quickly killed by liver failure with symptoms similar to fulminant hepatitis.[101,102] Thus, the FasL-independent activation of Fas offers some opportunities to find agents that can circumvent the above hepatic effects, but preserve Fas activating properties. Such a notion has found experimental support in our recent studies on the antitumor ether lipid ET-18-OCH₃.[45] This selective antitumor compound is incorporated in significant amounts only in tumor cells, and once inside the cell promotes apoptosis through intracellular activation and capping of Fas, independently of FasL[38,44,46] (Gajate et al, in preparation). Normal cells are spared because they do not take up the ether lipid.[38,44,45]

The increasing number of agents that promote FasL-independent activation of Fas through Fas clustering (Table 1) suggests that this process is more general than initially believed. The fact that very different experimental conditions and diverse agents, targeting distinct molecules and cellular processes, can lead eventually to an apoptotic response mediated by FasL-independent activation of Fas suggests that this process can be a general mechanism of cell death. Because most cells express Fas at their surface, we can hypothesize that when cells are committed to die, they can generate intracellular signals that trigger an efficient suicide mechanism from within the cell, via Fas activation, without receiving any order from outside through its ligand. Thus, Fas can receive efficiently orders either from outside, via FasL, or from inside the cell. Elucidation of the molecules and signaling pathways involved in this latter FasL-independent intracellular activation of Fas, through its capping into membrane rafts, is a major challenge for future research and can lead to identify new therapeutic targets.

Acknowledgments

Work from the authors' laboratory, described in this study, was supported by grants FIS-02/1199 and FIS-01/1048 from the Fondo de Investigación Sanitaria, grant SA-087/01 from Junta de Castilla y León, and grant 1FD97-0622 from the European Commission and Comisión Interministerial de Ciencia y Tecnología.

References

1. Yonehara S, Ishii A, Yonehara M. A cell-killing monoclonal antibody (anti-Fas) to a cell surface antigen codownregulated with the receptor of tumor necrosis factor. J Exp Med 1989; 169(5):1747-1756.
2. Yonehara S. To reviews on physiological and pathological roles of cell death. Cell Struct Funct 2003; 28(1):1-2.
3. Trauth BC, Klas C, Peters AM et al. Monoclonal antibody-mediated tumor regression by induction of apoptosis. Science 1989; 245(4915):301-305.
4. Itoh N, Yonehara S, Ishii A et al. The polypeptide encoded by the cDNA for human cell surface antigen Fas can mediate apoptosis. Cell 1991; 66(2):233-243.

5. Oehm A, Behrmann I, Falk W et al. Purification and molecular cloning of the APO-1 cell surface antigen, a member of the tumor necrosis factor/nerve growth factor receptor superfamily. Sequence identity with the Fas antigen. J Biol Chem 1992; 267(15):10709-10715.

6. Suda T, Takahashi T, Golstein P et al. Molecular cloning and expression of the Fas ligand, a novel member of the tumor necrosis factor family. Cell 1993; 75(6):1169-1178.

7. Takahashi T, Tanaka M, Brannan CI et al. Generalized lymphoproliferative disease in mice, caused by a point mutation in the Fas ligand. Cell 1994; 76(6):969-976.

8. Takahashi T, Tanaka M, Inazawa J et al. Human Fas ligand: Gene structure, chromosomal location and species specificity0 Int Immunol 1994; 6(10):1567-1574.

9. Nagata S. Apoptosis by death factor. Cell 1997; 88(3):355-365.

10. Tanaka M, Suda T, Takahashi T et al. Expression of the functional soluble form of human fas ligand in activated lymphocytes. Embo J 1995; 14(6):1129-1135.

11. Itoh N, Nagata S. A novel protein domain required for apoptosis. Mutational analysis of human Fas antigen. J Biol Chem 1993; 268(15):10932-10937.

12. Huang B, Eberstadt M, Olejniczak ET et al. NMR structure and mutagenesis of the Fas (APO-1/CD95) death domain. Nature 1996; 384(6610):638-641.

13. Liang H, Fesik SW. Three-dimensional structures of proteins involved in programmed cell death. J Mol Biol 1997; 274(3):291-302.

14. Siegel RM, Chan FK, Chun HJ et al. The multifaceted role of Fas signaling in immune cell homeostasis and autoimmunity. Nat Immunol 2000; 1(6):469-474.

15. Chan FK, Chun HJ, Zheng L et al. A domain in TNF receptors that mediates ligand-independent receptor assembly and signaling. Science 2000; 288(5475):2351-2354.

16. Papoff G, Hausler P, Eramo A et al. Identification and characterization of a ligand-independent oligomerization domain in the extracellular region of the CD95 death receptor. J Biol Chem 1999; 274(53):38241-38250.

17. Siegel RM, Frederiksen JK, Zacharias DA et al. Fas preassociation required for apoptosis signaling and dominant inhibition by pathogenic mutations. Science 2000; 288(5475):2354-2357.

18. Chinnaiyan AM, O'Rourke K, Tewari M et al. FADD, a novel death domain-containing protein, interacts with the death domain of Fas and initiates apoptosis. Cell 1995; 81(4):505-512.

19. Boldin MP, Goncharov TM, Goltsev YV et al. Involvement of MACH, a novel MORT1/FADD-interacting protease, in Fas/APO-1- and TNF receptor-induced cell death. Cell 1996; 85(6):803-815.

20. Ashkenazi A, Dixit VM. Death receptors: Signaling and modulation. Science 1998; 281(5381):1305-1308.

21. Kischkel FC, Hellbardt S, Behrmann I et al. Cytotoxicity-dependent APO-1 (Fas/CD95)-associated proteins form a death-inducing signaling complex (DISC) with the receptor. Embo J 1995; 14(22):5579-5588.

22. Watanabe-Fukunaga R, Brannan CI, Copeland NG et al. Lymphoproliferation disorder in mice explained by defects in Fas antigen that mediates apoptosis. Nature 1992; 356(6367):314-317.

23. Algeciras-Schimnich A, Shen L, Barnhart BC et al. Molecular ordering of the initial signaling events of CD95. Mol Cell Biol 2002; 22(1):207-220.

24. Song HY, Dunbar JD, Donner DB. Aggregation of the intracellular domain of the type 1 tumor necrosis factor receptor defined by the two-hybrid system. J Biol Chem 1994; 269(36):22492-22495.

25. Boldin MP, Mett IL, Varfolomeev EE et al. Self-association of the "death domains" of the p55 tumor necrosis factor (TNF) receptor and Fas/APO1 prompts signaling for TNF and Fas/APO1 effects. J Biol Chem 1995; 270(1):387-391.

26. Friesen C, Herr I, Krammer PH et al. Involvement of the CD95 (APO-1/FAS) receptor/ligand system in drug- induced apoptosis in leukemia cells. Nat Med 1996; 2(5):574-577.

27. Muller M, Strand S, Hug H et al. Drug-induced apoptosis in hepatoma cells is mediated by the CD95 (APO-1/Fas) receptor/ligand system and involves activation of wild-type p53. J Clin Invest 1997; 99(3):403-413.

28. Fulda S, Sieverts H, Friesen C et al. The CD95 (APO-1/Fas) system mediates drug-induced apoptosis in neuroblastoma cells. Cancer Res 1997; 57(17):3823-3829.

29. Poulaki V, Mitsiades CS, Mitsiades N. The role of Fas and FasL as mediators of anticancer chemotherapy. Drug Resist Updat 2001; 4(4):233-242.

30. Fulda S, Strauss G, Meyer E et al. Functional CD95 ligand and CD95 death-inducing signaling complex in activation-induced cell death and doxorubicin-induced apoptosis in leukemic T cells. Blood 2000; 95(1):301-308.

31. Nagarkatti N, Davis BA. Tamoxifen induces apoptosis in Fas+ tumor cells by upregulating the expression of Fas ligand. Cancer Chemother Pharmacol 2003; 51(4):284-290.

32. Villunger A, Egle A, Kos M et al. Drug-induced apoptosis is associated with enhanced Fas (Apo-1/CD95) ligand expression but occurs independently of Fas (Apo-1/CD95) signaling in human T-acute lymphatic leukemia cells. Cancer Res 1997; 57(16):3331-3334.

33. Gamen S, Anel A, Lasierra P et al. Doxorubicin-induced apoptosis in human T-cell leukemia is mediated by caspase-3 activation in a Fas-independent way. FEBS Lett 1997; 417(3):360-364.

34. Eischen CM, Kottke TJ, Martins LM et al. Comparison of apoptosis in wild-type and Fas-resistant cells: Chemotherapy-induced apoptosis is not dependent on Fas/Fas ligand interactions. Blood 1997; 90(3):935-943.

35. Tolomeo M, Dusonchet L, Meli M et al. The CD95/CD95 ligand system is not the major effector in anticancer drug-mediated apoptosis. Cell Death Differ 1998; 5(9):735-742.

36. Wesselborg S, Engels IH, Rossmann E et al. Anticancer drugs induce caspase-8/FLICE activation and apoptosis in the absence of CD95 receptor/ligand interaction. Blood 1999; 93(9):3053-3063.

37. Ferreira CG, Tolis C, Span SW et al. Drug-induced apoptosis in lung cnacer cells is not mediated by the Fas/FasL (CD95/APO1) signaling pathway. Clin Cancer Res 2000; 6(1):203-212.

38. Gajate C, Fonteriz RI, Cabaner C et al. Intracellular triggering of Fas, independently of FasL, as a new mechanism of antitumor ether lipid-induced apoptosis. Int J Cancer 2000; 85(5):674-682.

39. Fulda S, Friesen C, Los M et al. Betulinic acid triggers CD95 (APO-1/Fas)- and p53-independent apoptosis via activation of caspases in neuroectodermal tumors. Cancer Res 1997; 57(21):4956-4964.

40. Newton K, Strasser A. Ionizing radiation and chemotherapeutic drugs induce apoptosis in lymphocytes in the absence of Fas or FADD/MORT1 signaling. Implications for cancer therapy. J Exp Med 2000; 191(1):195-200.

41. Micheau O, Solary E, Hammann A et al. Fas ligand-independent, FADD-mediated activation of the Fas death pathway by anticancer drugs. J Biol Chem 1999; 274(12):7987-7992.

42. Matzke A, Massing U, Krug HF. Killing tumour cells by alkylphosphocholines: Evidence for involvement of CD95. Eur J Cell Biol 2001; 80(1):1-10.

43. Bertin J, Armstrong RC, Ottilie S et al. Death effector domain-containing herpesvirus and poxvirus proteins inhibit both Fas- and TNFR1-induced apoptosis. Proc Natl Acad Sci USA 1997; 94(4):1172-1176.

44. Mollinedo F, Fernandez-Luna JL, Gajate C et al. Selective induction of apoptosis in cancer cells by the ether lipid ET- 18-OCH3 (Edelfosine): Molecular structure requirements, cellular uptake, and protection by Bcl-2 and Bcl-X(L). Cancer Res 1997; 57(7):1320-1328.

45. Gajate C, Mollinedo F. Biological Activities, Mechanisms of Action and Biomedical Prospect of the Antitumor Ether Phospholipid ET-18-OCH(3) (Edelfosine), A Proapoptotic Agent in Tumor Cells. Curr Drug Metab 2002; 3(5):491-525.

46. Gajate C, Mollinedo F. The antitumor ether lipid ET-18-OCH(3) induces apoptosis through translocation and capping of Fas/CD95 into membrane rafts in human leukemic cells. Blood 2001; 98(13):3860-3863.

47. Hueber AO, Bernard AM, Herincs Z et al. An essential role for membrane rafts in the initiation of Fas/CD95-triggered cell death in mouse thymocytes. EMBO Rep 2002; 3(2):190-196.

48. Scheel-Toellner D, Wang K, Singh R et al. The death-inducing signalling complex is recruited to lipid rafts in Fas-induced apoptosis. Biochem Biophys Res Commun 2002; 297(4):876-879.

49. Delmas D, Rebe C, Lacour S et al. Resveratrol-induced apoptosis is associated with Fas redistribution in the rafts and the formation of a death-inducing signaling complex in colon cancer cells. J Biol Chem 2003; 278(42):41482-41490.

50. Simons K, van Meer G. Lipid sorting in epithelial cells. Biochemistry 1988; 27(17):6197-6202.

51. Simons K, Toomre D. Lipid rafts and signal transduction. Nat Rev Mol Cell Biol 2000; 1(1):31-39.

52. Ikonen E. Roles of lipid rafts in membrane transport. Curr Opin Cell Biol 2001; 13(4):470-477.

53. Nichols BJ, Lippincott-Schwartz J. Endocytosis without clathrin coats. Trends Cell Biol 2001; 11(10):406-412.

54. Manes S, del Real G, Martinez AC. Pathogens: Raft hijackers. Nat Rev Immunol 2003; 3(7):557-568.
55. Brown DA, London E. Structure and function of sphingolipid- and cholesterol-rich membrane rafts. J Biol Chem 2000; 275(23):17221-17224.
56. Grassme H, Jekle A, Riehle A et al. CD95 signaling via ceramide-rich membrane rafts. J Biol Chem 2001; 276(23):20589-20596.
57. Grassme H, Cremesti A, Kolesnick R et al. Ceramide-mediated clustering is required for CD95-DISC formation. Oncogene 2003; 22(35):5457-5470.
58. Cremesti A, Paris F, Grassme H et al. Ceramide enables fas to cap and kill. J Biol Chem 2001; 276(26):23954-23961.
59. Grassme H, Schwarz H, Gulbins E. Molecular mechanisms of ceramide-mediated CD95 clustering. Biochem Biophys Res Commun 2001; 284(4):1016-1030.
60. Kolesnick R. The therapeutic potential of modulating the ceramide/sphingomyelin pathway. J Clin Invest 2002; 110(1):3-8.
61. Prinetti A, Chigorno V, Prioni S et al. Changes in the lipid turnover, composition, and organization, as sphingolipid-enriched membrane domains, in rat cerebellar granule cells developing in vitro. J Biol Chem 2001; 276(24):21136-21145.
62. Cremesti AE, Goni FM, Kolesnick R. Role of sphingomyelinase and ceramide in modulating rafts: Do biophysical properties determine biologic outcome? FEBS Lett 2002; 531(1):47-53.
63. Debatin KM, Beltinger C, Bohler T et al. Regulation of apoptosis through CD95 (APO-I/Fas) receptor-ligand interaction. Biochem Soc Trans 1997; 25(2):405-410.
64. Cabaner C, Gajate C, Macho A et al. Induction of apoptosis in human mitogen-activated peripheral blood T- lymphocytes by the ether phospholipid ET-18-OCH3: involvement of the Fas receptor/ligand system. Br J Pharmacol 1999; 127(4):813-825.
65. Rehemtulla A, Hamilton CA, Chinnaiyan AM et al. Ultraviolet radiation-induced apoptosis is mediated by activation of CD-95 (Fas/APO-1). J Biol Chem 1997; 272(41):25783-25786.
66. Bennett M, Macdonald K, Chan SW et al. Cell surface trafficking of Fas: A rapid mechanism of p53-mediated apoptosis. Science 1998; 282(5387):290-293.
67. Aragane Y, Kulms D, Metze D et al. Ultraviolet light induces apoptosis via direct activation of CD95 (Fas/APO-1) independently of its ligand CD95L. J Cell Biol 1998; 140(1):171-182.
68. Tang D, Lahti JM, Grenet J et al. Cycloheximide-induced T-cell death is mediated by a Fas-associated death domain-dependent mechanism. J Biol Chem 1999; 274(11):7245-7252.
69. Chen Y, Lai MZ. c-Jun NH2-terminal kinase activation leads to a FADD-dependent but Fas ligand-independent cell death in Jurkat T cells. J Biol Chem 2001; 276(11):8350-8357.
70. Beltinger C, Fulda S, Kammertoens T et al. Herpes simplex virus thymidine kinase/ganciclovir-induced apoptosis involves ligand-independent death receptor aggregation and activation of caspases. Proc Natl Acad Sci USA 1999; 96(15):8699-8704.
71. Bush JA, Cheung Jr KJ, Li G. Curcumin induces apoptosis in human melanoma cells through a Fas receptor/caspase-8 pathway independent of p53. Exp Cell Res 2001; 271(2):305-314.
72. Luo J, Sun Y, Lin H et al. Activation of JNK by vanadate induces a Fas-associated death domain (FADD)-dependent death of cerebellar granule progenitors in vitro. J Biol Chem 2003; 278(7):4542-4551.
73. Muzio M, Stockwell BR, Stennicke HR et al. An induced proximity model for caspase-8 activation. J Biol Chem 1998; 273(5):2926-2930.
74. Martin DA, Siegel RM, Zheng L et al. Membrane oligomerization and cleavage activates the caspase-8 (FLICE/MACHalpha1) death signal. J Biol Chem 1998; 273(8):4345-4349.
75. Siegel RM, Martin DA, Zheng L et al. Death-effector filaments: Novel cytoplasmic structures that recruit caspases and trigger apoptosis. J Cell Biol 1998; 141(5):1243-1253.
76. Perez D, White E. E1B 19K inhibits Fas-mediated apoptosis through FADD-dependent sequestration of FLICE. J Cell Biol 1998; 141(5):1255-1266.
77. Goncalves A, Braguer D, Carles G et al. Caspase-8 activation independent of CD95/CD95-L interaction during paclitaxel-induced apoptosis in human colon cancer cells (HT29-D4). Biochem Pharmacol 2000; 60(11):1579-1584.
78. Bonny C, Oberson A, Negri S et al. Cell-permeable peptide inhibitors of JNK: novel blockers of beta-cell death. Diabetes 2001; 50(1):77-82.

79. Gajate C, Santos-Beneit A, Modolell M et al. Involvement of c-Jun NH2-terminal kinase activation and c-Jun in the induction of apoptosis by the ether phospholipid 1-O-octadecyl-2-O-methyl-rac-glycero-3-phosphocholine. Mol Pharmacol 1998; 53(4):602-612.

80. Verheij M, Bose R, Lin XH et al. Requirement for ceramide-initiated SAPK/JNK signalling in stress- induced apoptosis. Nature 1996; 380(6569):75-79.

81. Chen YR, Wang X, Templeton D et al. The role of c-Jun N-terminal kinase (JNK) in apoptosis induced by ultraviolet C and gamma radiation. Duration of JNK activation may determine cell death and proliferation. J Biol Chem 1996; 271(50):31929-31936.

82. Chen YR, Zhou G, Tan TH. c-Jun N-terminal kinase mediates apoptotic signaling induced by N-(4- hydroxyphenyl)retinamide. Mol Pharmacol 1999; 56(6):1271-1279.

83. Gajate C, An F, Mollinedo F. Rapid and Selective Apoptosis in Human Leukemic Cells Induced by Aplidine through a Fas/CD95- and Mitochondrial-mediated Mechanism. Clin Cancer Res 2003; 9(4):1535-1545.

84. Gajate C, An F, Mollinedo F. Differential cytostatic and apoptotic effects of ecteinascidin-743 in cancer cells. Transcription-dependent cell cycle arrest and transcription-independent JNK and mitochondrial mediated apoptosis. J Biol Chem 2002; 277(44):41580-41589.

85. Yang DD, Kuan CY, Whitmarsh AJ et al. Absence of excitotoxicity-induced apoptosis in the hippocampus of mice lacking the Jnk3 gene. Nature 1997; 389(6653):865-870.

86. Kuan CY, Yang DD, Samanta Roy DR et al. The Jnk1 and Jnk2 protein kinases are required for regional specific apoptosis during early brain development. Neuron 1999; 22(4):667-676.

87. Yamamoto K, Ichijo H, Korsmeyer SJ. BCL-2 is phosphorylated and inactivated by an ASK1/Jun N-terminal protein kinase pathway normally activated at G(2)/M. Mol Cell Biol 1999; 19(12):8469-8478.

88. Kharbanda S, Saxena S, Yoshida K et al. Translocation of SAPK/JNK to mitochondria and interaction with Bcl-x(L) in response to DNA damage. J Biol Chem 2000; 275(1):322-327.

89. Tournier C, Hess P, Yang DD et al. Requirement of JNK for stress-induced activation of the cytochrome c- mediated death pathway. Science 2000; 288(5467):870-874.

90. Mollinedo F, Gajate C. Microtubules, microtubule-interfering agents and apoptosis. Apoptosis 2003; 8(5):413-450.

91. Luna EJ, Hitt AL. Cytoskeleton—plasma membrane interactions. Science 1992; 258(5084):955-964.

92. Tsukita S, Oishi K, Sato N et al. ERM family members as molecular linkers between the cell surface glycoprotein CD44 and actin-based cytoskeletons. J Cell Biol 1994; 126(2):391-401.

93. Dransfield DT, Bradford AJ, Smith J et al. Ezrin is a cyclic AMP-dependent protein kinase anchoring protein. Embo J 1997; 16(1):35-43.

94. Harder T, Simons K. Clusters of glycolipid and glycosylphosphatidylinositol-anchored proteins in lymphoid cells: Accumulation of actin regulated by local tyrosine phosphorylation. Eur J Immunol 1999; 29(2):556-562.

95. Bretscher A. Regulation of cortical structure by the ezrin-radixin-moesin protein family. Curr Opin Cell Biol 1999; 11(1):109-116.

96. Mangeat P, Roy C, Martin M. ERM proteins in cell adhesion and membrane dynamics. Trends Cell Biol 1999; 9(5):187-192.

97. Parlato S, Giammarioli AM, Logozzi M et al. CD95 (APO-1/Fas) linkage to the actin cytoskeleton through ezrin in human T lymphocytes: A novel regulatory mechanism of the CD95 apoptotic pathway. EMBO J 2 2000; 19(19):5123-5134.

98. Giammarioli AM, Garofalo T, Sorice M et al. GD3 glycosphingolipid contributes to Fas-mediated apoptosis via association with ezrin cytoskeletal protein. FEBS Lett 2001; 506(1):45-50.

99. Yujiri T, Fanger GR, Garrington TP et al. MEK kinase 1 (MEKK1) transduces c-Jun NH2-terminal kinase activation in response to changes in the microtubule cytoskeleton. J Biol Chem 1999; 274(18):12605-12610.

100. Gajate C, Barasoain I, Andreu JM et al. Induction of apoptosis in leukemic cells by the reversible microtubule- disrupting agent 2-methoxy-5-(2',3',4'-trimethoxyphenyl)-2,4,6- cycloheptatrien-1 -one: Protection by Bcl-2 and Bcl-X(L) and cell cycle arrest. Cancer Res 2000; 60(10):2651-2659.

101. Ogasawara J, Watanabe-Fukunaga R, Adachi M et al. Lethal effect of the anti-Fas antibody in mice. Nature 1993; 364(6440):806-809.

102. Tanaka M, Suda T, Yatomi T et al. Lethal effect of recombinant human Fas ligand in mice pretreated with Propionibacterium acnes. J Immunol 1997; 158(5):2303-2309.

103. Shao RG, Cao CX, Nieves-Neira W et al. Activation of the Fas pathway independently of Fas ligand during apoptosis induced by camptothecin in p53 mutant human colon carcinoma cells. Oncogene 2001; 20(15):1852-1859.

104. Fumarola C, Zerbini A, Guidotti GG. Glutamine deprivation-mediated cell shrinkage induces ligand-independent CD95 receptor signaling and apoptosis. Cell Death Differ 2001; 8(10):1004-1013.

105. Moorman JP, Prayther D, McVay D et al. The C-terminal region of hepatitis C core protein is required for Fas-ligand independent apoptosis in Jurkat cells by facilitating Fas oligomerization. Virology 2003; 312(2):320-329.

106. Kim SG, Jong HS, Kim TY et al. Transforming Growth Factor-{beta}1 Induces Apoptosis through Fas Ligand-independent Activation of the Fas Death Pathway in Human Gastric SNU-620 Carcinoma Cells. Mol Biol Cell 2003.

107. Zhuang S, Kochevar IE. Ultraviolet A radiation induces rapid apoptosis of human leukemia cells by Fas ligand-independent activation of the Fas death pathways. Photochem Photobiol 2003; 78(1):61-67.

CHAPTER 3

Role of Ceramide in CD95 Signaling

Volker Teichgräber, Gabriele Hessler and Erich Gulbins

Abstract

Recent studies indicate that the reorganization of receptor molecules in distinct domains of the cell membrane constitutes an important and general mechanism that is required for the initiation of signaling via various receptor molecules. Studies on the CD95 receptor might serve as a paradigm for the mechanism mediating receptor clustering/aggregation. These studies revealed activation of the acid sphingomyelinase and a release of ceramide in the outer leaflet of the cell membrane upon stimulation of CD95. The unique biophysical properties of ceramide trigger the formation of large ceramide-enriched membrane platforms that serve to trap and cluster the receptor. This process results in a high density of CD95 within a distinct area of the cell membrane and amplifies the primary signal generated by binding of the CD95 ligand and, thus, permits the induction of apoptosis. Furthermore, ceramide-enriched membrane domains mediate the assembling of the receptor with intracellular signaling molecules, in particular FADD, caspase 8, and the potassium channel Kv1.3 that finally mediate apoptosis initiated by CD95 ligation.

Membrane Rafts

Biological membranes of eukaryotic cells are primarily composed of glycerophospholipids, sphingolipids and cholesterol. The classical fluid mosaic model of Singer and Nicolson (1972) that suggested a mixture of the lipids in the cell membrane[1] was modified in recent years to accommodate the formation of distinct lipid microdomains within the cell membrane.[2,3] These distinct domains, which are intimately involved in signaling functions, are mainly composed of sphingolipids and cholesterol. In detail, the headgroups of sphingolipids associate via hydrophilic interactions that are enhanced by hydrophobic interactions between the saturated side chains of the sphingolipids. Since the headgroups are rather bulky, the lateral organization of sphingolipids needs to be stabilized by a molecule that bridges void spaces between the large glycerosphingolipids. This function is mediated by cholesterol that tightly interacts with sphingolipids, in particular sphingomyelin, by hydrophilic and hydrophobic bonding through the C3-hydroxyl group and the sterolring system with the sphingosine moiety of sphingomyelin, respectively.

The tight interaction of sphingolipids with one another and with cholesterol results in the segregation of these lipids from other glycerophospholipids in the cell membrane and, thus, in a lateral organization into discrete membrane structures characterized by a liquid-ordered or even gel-like phase.[2,3] These distinct sphingolipid- and cholesterol-enriched membrane microdomains that are relatively resistant to detergents have been termed detergent-insoluble

Fas Signaling, edited by Harald Wajant. ©2006 Landes Bioscience and Springer Science+Business Media.

glycolipid enriched membranes (DIGs)[4,5] or rafts since they are considered to float in an "ocean" of other membrane phospholipids.[2]

Although many studies suggest the existence of rafts in the cell membrane, the exact nature of rafts is still controversial. In particular, it has been suggested that the extraction of the cell membrane with detergents does not permit to accurately determine the nature of rafts.[6] In addition, since rafts are very small, a direct visualization of rafts on living cells still has to be achieved. Finally, it appears that distinct domains formed by cholesterol and sphingolipids cover a large part of the plasma membrane,[7] an observation that does not fit with the concept of small rafts floating in the cell membrane. Therefore, further techniques have to be developed to further characterize and directly prove the existence of small, primary membrane rafts.

Ceramide-Enriched Membrane Platforms

The most prevalent sphingolipid in rafts is sphingomyelin that consists of a highly hydrophobic ceramide moiety and a hydrophilic phosphorylcholine headgroup. Ceramide is the amide ester of the sphingoid base D-erythro-sphingosine and a fatty acid usually of C_{16} through C_{26} chain length.

Ceramide is generated in cells upon rapid hydrolysis of plasma membrane sphingomyelin by the activity of a sphingomyelinase or by de novo synthesis regulated by the enzymes serine-palmitoyl-transferase and ceramide synthase. Three distinct forms of sphingomyelinases have been identified in mammalian cells at present and are discriminated by their pH optimum, i.e., the acid, neutral and alkaline sphingomyelinase.[8-10] In the present overview, we would like to focus on the function of the acid sphingomyelinase (ASM).

In the recent years, we have demonstrated a novel mechanism how ceramide may function in cells and how it may form a membrane signaling unit.[11-15]

Our studies demonstrated that several receptor molecules, e.g., CD95 or CD40, activate the acid sphingomyelinase and cause the appearance of the enzyme on the extracellular leaflet of the plasma membrane (Figs. 1, 2).[11,12,14] Neither the exact mechanisms of the activation of the acid sphingomyelinase nor those of the surface exposure of the protein are known. Recent data from Qui et al suggest that the acid sphingomyelinase is regulated in vitro by oxidation of a specific cysteine.[16] However, it still has to be determined whether a redox mechanism does also apply to the regulation of the acid sphingomyelinase in vivo. Further studies indicated a regulation of the acid sphingomyelinase by phosphatidylinositol-3,5-bis-phosphate.[17] Whether this mechanism is operative in vivo is also unknown. Although not proven at present, some preliminary data from our laboratory suggest that the acid sphingomyelinase in resting cells is contained within secretory vesicles that are very rapidly mobilized upon appropriate stimulation, e.g., CD95 ligation. These vesicles fuse with the cell membrane and expose their content including the acid sphingomyelinase to the extracellular space. The activity of the acid sphingomyelinase in the extracellular leaflet of the cell membrane that also contains the majority of cellular sphingomyelin,[18] results in the generation of ceramide and in the formation of large, ceramide-enriched membrane platforms[11,12,14] that are most likely generated by a lateral coalescence of microdomains.[3] The formation of large ceramide-enriched membrane platforms is most likely mediated by the biophysical properties of ceramide to spontaneously self-aggregate to domains that fuse.[3,19,20] The in vivo data provided by our group are supported by in vitro studies that demonstrated the formation of a ceramide-rich macrodomain in phosphatidyl-choline/sphingomyelin-composed unilamellar vesicles, which were locally incubated with a sphingomyelinase.[21] Further in vitro magnetic resonance spectroscopy and atomic force microscopy studies showed that addition of ceramide is sufficient to transform a fluid phospholipid bilayer into a gel phase.[22] Second, ceramide partitions into distinct domains in model membranes and does not mix with other phospholipids[23] and, third, long chain ceramides undergo lateral phase separation in glycerophospholipid/cholesterol bilayers.[24]

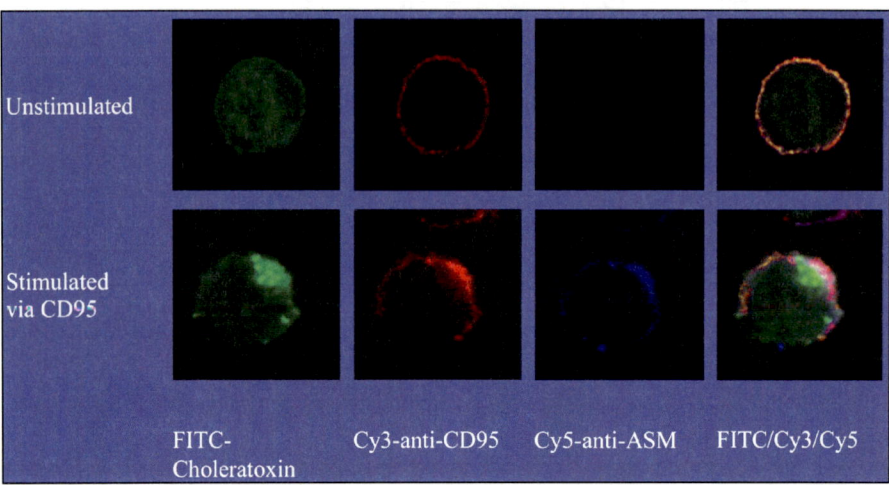

| | FITC-Choleratoxin | Cy3-anti-CD95 | Cy5-anti-ASM | FITC/Cy3/Cy5 |

Figure 1. CD95 and acid sphingomyelinase colocalize in membrane platforms. JY B lymphocytes were stimulated via CD95 for 2 min, fixed and stained with Cy3-coupled anti-CD95 and Cy5-labeled anti-acid sphingomyelinase antibodies. Membrane rafts were visualized by staining with FITC-coupled choleratoxin that binds to gangliosides enriched in membrane rafts. CD95 was homogenously distributed in the cell membrane prior to activation and very rapidly formed a cluster after stimulation. The acid sphingomyelinase was almost absent from the surface of unstimulated cells, while stimulation resulted in exposure and clustering of the protein on the cell surface. The FITC-choleratoxin staining reveals profound reorganization of membrane rafts to a membrane platform that contains clustered CD95 and surface acid sphingomyelinase, as evidenced in the overlay picture. (Printed with permission of *J Biol Chem.*)

Although we assume that ceramide is generated within rafts, the formation of ceramide-enriched membrane domains and signaling via these macrodomains does not necessarily require the presence of small membrane rafts, since the tendency of ceramide molecules to self-associate and to separate from phospholipids already seems to be sufficient for the generation of ceramide-enriched membrane platforms.

The maximum activity of the acid sphingomyelinase has been shown to be at acidic pH.[25] However, a rise of the pH to neutral values as they exist at the cell surface only increases the K_m value of the acid sphingomyelinase, but does not alter the activity [V_{max}] of the enzyme.[26] Therefore, modifications of the environmental conditions that promote substrate affinity of the acid sphingomyelinase are capable to restore enzyme activity on the cell surface. Studies from Tabas et al were able to identify those factors that include LDL and further membrane cofactors.[14]

Function of Ceramide-Enriched Membrane Platforms for CD95 Signaling

Receptor Clustering

At present, we favor the concept that ceramide-enriched membrane domains function as a sorting device in the cell membrane. These macro-domains trigger a reorganization of receptor molecules and intracellular signaling molecules, which finally initiates and mediates signal transduction into the cell. Ceramide-enriched membrane platforms have been shown to trap and cluster CD95.[11-13] Clustering or aggregation of CD95 in ceramide-enriched membrane platforms may have several functions: First, clustering results in a high local density of the

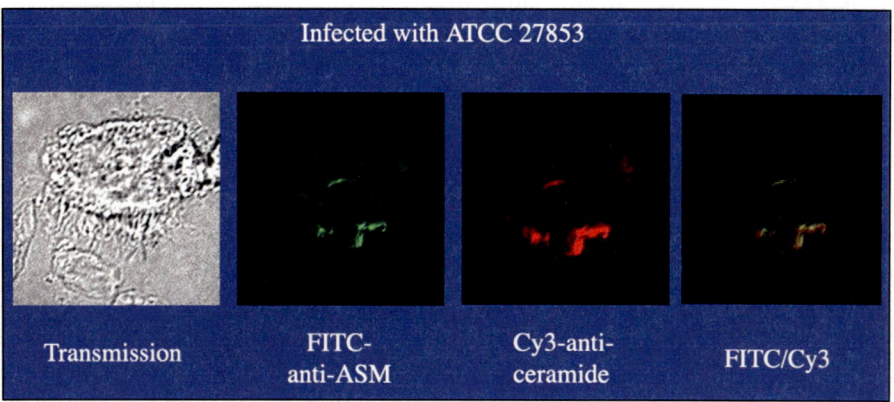

Figure 2. Ceramide-enriched membrane platforms are formed on the surface of tracheal epithelial cells upon infection with *P. aeruginosa*. In situ infection of murine tracheal epithelial cells with *P. aeruginosa* results in the formation of large ceramide-enriched membrane platforms that contain surface acid sphingomyelinase. Uninfected epithelial cells did not display any ceramide or acid sphingomyelinase on the cell surface (not shown). Ceramide and acid sphingomyelinase were visualized by staining epithelial cells, which had been isolated from the trachea 20 min after infection, with FITC-labeled anti-acid sphingomyelinase and Cy3-coupled anti-ceramide antibodies. Samples were analyzed by fluorescence microscopy. (Printed with permission of Nature Publishing Group.)

receptor in a distinct domain of the cell membrane and promotes oligomerization of the receptor that seems to be required for signaling via CD95.[27] This was demonstrated in cells lacking the acid sphingomyelinase, which are unable to release ceramide upon CD95 activation.[27] Stimulation of those cells via CD95 resulted in a very weak, but detectable recruitment of FADD to the receptor molecule and a slight activation of caspase 8, while downstream effectors, e.g., release of cytochrome c from mitochondria, activation of caspase 3 or induction of apoptosis were absent. Addition of natural C_{16}-ceramide to the cells restored CD95 clustering and permitted complete activation of caspase 8, caspase 3 and execution of apoptosis. Thus, receptor clustering serves to strongly amplify a primary weak signal. As for CD95 we observed an approximately 100-fold amplification of the primary stimulus by clustering.[27]

The rapid aggregation/clustering of CD95 upon stimulation was demonstrated to occur in many cell types including lymphocytes, phagocytic cells, granulosa cells of the ovary, epithelial cells, fibroblasts, hepatocytes, and thymocytes and seems to constitute a general mechanism of CD95 signaling.[28]

These data indicate a central role of the acid sphingomyelinase in CD95 signaling, which was also elaborated in other systems. It was demonstrated that human and murine lympho-cytes[29] or hepatocytes,[29,30] respectively, that are deficient for the acid sphingomyelinase, are resistant to CD95-induced cell death both in vitro and in vivo. These cells were derived from a Niemann-Pick disease type A patient, suffering from an inborne deficiency of ASM or from mice genetically deficient for the acid sphingomyelinase. Injection of anti-CD95 antibodies into those mice failed to induce apoptosis in liver cells and to cause death of the animal, events that were rapidly observed in acid sphingomyelinase-positive control mice.[29,30] Retransfection of the cells with an expression vector for the acid sphingomyelinase or treatment of ASM-deficient cells with natural C_{16}-ceramide restored CD95-triggered apoptosis, indicating that ceramide is required for CD95-triggered apoptosis. Likewise, acid sphingomyelinase-deficient cells or mice, respectively, were resistant to the induction of apoptosis by tumor necrosis factor[31] that has been previously shown to activate the acid sphingomyelinase and to induce a release of ceramide.[32] Furthermore, two additional forms of CD95-mediated apoptosis in vivo are abrogated or at least blunted in mice lacking the acid sphingomyelinase: Injection of phytohemagglutinin into mice (PHA) induces expression of CD95 ligand on lymphocytes that migrate to the liver and induce apoptosis in CD95-positive hepatocytes.[33] This approach results in an autoimmune hepatitis in wild-type mice, while ASM-deficient mice were protected against the adverse effects of PHA injections.[29] Second, the ligation of CD4 on T cells upon intravenous injection of anti-CD4 antibodies results in a CD95 and CD95 ligand-dependent apoptosis of lymphocytes.[34] In contrast, CD4-positive T cells in acid sphingomyelinase-deficient mice were protected from apoptosis upon injection of anti-CD4 antibodies,[29] indicating the in vivo significance of the acid sphingomyelinase for apoptosis.

Different cell types are developmentally endowed with distinct mechanisms to cluster CD95 upon stimulation. For instance, Jurkat cells do not display CD95 in rafts prior to stimulation, but trap the receptor in ceramide-enriched membrane platforms that are formed after CD95 stimulation.[35] Other cells, in particular thymocytes, display a constitutive localization of CD95 in rafts.[36] The formation of ceramide in rafts of those cells immediately results in clustering of the receptor without the requirement for trapping the receptor molecule beforehand. At present, the mechanisms determining the presence or absence, respectively, of CD95 in rafts of unstimulated cells are unknown.

Reorganization of Intracellular Signaling Molecules

Ceramide-enriched membrane rafts might be also involved in the reorganization of intracellular signaling molecules that mediate the effect of CD95 or other receptors clustering in ceramide-enriched membrane domains. Thus, it has been shown that FADD and caspase 8 are recruited to membrane rafts upon stimulation of CD95.[37,38] Destruction of rafts prevented induction of apoptosis via the FADD/caspase 8 pathway in several cell systems.[37,38] In addition, ceramide-enriched membrane domains may serve to exclude certain molecules from signaling platforms that could counteract CD95 and prevent apoptosis. This concept is consistent with the recent finding that caspase 8 is recruited to rafts upon CD95 stimulation and associates with the raft ganglioside GM3.[37] Destruction of rafts by extraction of cholesterol prevented CD95-induced cell death. These data emphasize the significance of rafts and the assembly of signaling molecules within those structures for CD95-indced cell death. Furthermore, the drug resveratrol induced clustering, reorganization and association of CD95, the adapter protein FADD, and caspase 8 with cholesterol- and sphingolipid-enriched rafts.[39] Finally, we have recently shown that the Kv1.3 ion channel localizes into ceramide-enriched membrane domains[35] and is inactivated upon cellular stimulation via CD95.[40] Since this channel has been reported to be required for actinomycin-induced cell death,[41] it might be an additional component of the signaling machinery assembled in ceramide-enriched membrane platforms to trigger CD95 induced apoptosis.

A very similar concept of reorganization of signaling molecules has been recently demonstrated for the TCR/CD3-complex that also localizes to rafts.[42] Activation of the cells via the TCR/CD3 complex resulted in a dramatic reorganization of these proteins and in exclusion of approximately 50% of all proteins from rafts.[42] However, it is unknown whether ceramide-enriched membrane platforms play a role in T-cell receptor signaling.

A recent manuscript indicated that ceramide displaces cholesterol from membrane rafts.[43] Therefore, the generation of ceramide within rafts might dramatically alter the biophysical properties of these rafts and may result in the exclusion of molecules that preferentially associate with cholesterol. These studies provide a novel mechanism for the reorganization of signaling molecules within rafts upon generation of ceramide.

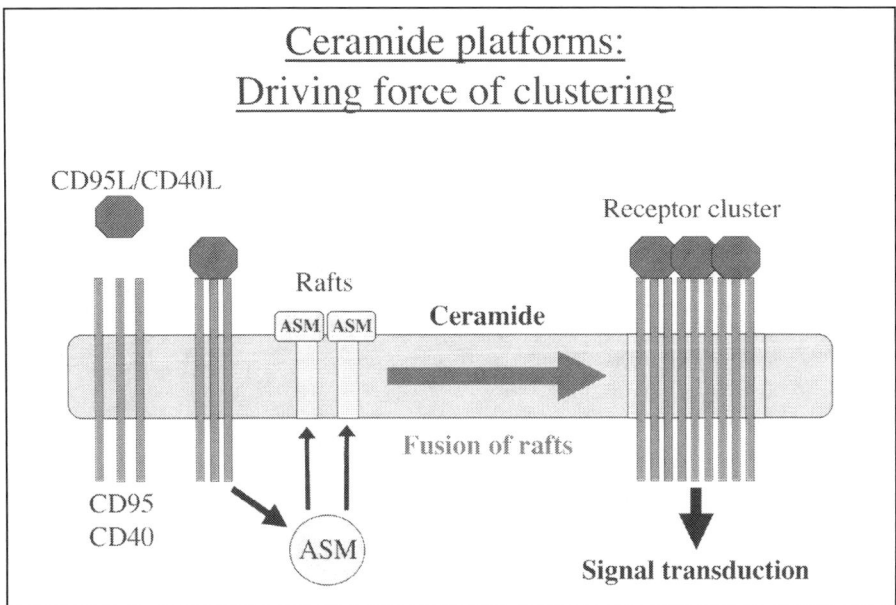

Figure 3. Model of the formation and function of ceramide-enriched membrane platforms in CD95 signaling. Ligation of CD95 or CD40 induces activation and translocation of the acid sphingomyelinase (ASM) onto the extracellular leaflet of the cell membrane. The acid sphingomyelinase catalyzes the release of ceramide from sphingomyelin in the outer leaflet of the cell membrane. Ceramide then mediates reorganization of small membrane rafts into large ceramide-enriched membrane platforms that serve to cluster CD95 or CD40, to recruit intracellular signaling molecules and/or to exclude inhibitory molecules from the activated receptors. Therefore, it is these platforms that finally mediate apoptosis by CD95 or differentiation/proliferation by CD40.

Regulation of Receptor-Ligand Interactions

Finally, ceramide-enriched membrane platforms may stabilize the interaction of the receptor with the ligand and/or may directly alter the affinity/avidity of the receptor for its ligand. Whether such an alteration of the receptor-ligand interactions is simply mediated by the stabilization of the contact between the receptor-ligand pair or even promoted by ceramide-induced conformational changes of the receptor is presently unknown.

Mechanisms of Receptor Clustering

The mechanisms of receptor trapping in ceramide-enriched membrane rafts are largely unknown. Studies from our laboratory addressed the effect of an exchange of the transmembranous domain of CD40 with that of CD45 on the ability of the fusion protein to cluster in ceramide-enriched membrane platforms.[44] CD40 has been previously shown to cluster in ceramide-enriched membrane domains,[14] while CD45 does not.[45] The data revelead that the exchange of the transmembranous domain in the CD40 molecule prevented its clustering in ceramide-enriched membrane platforms indicating that the structure of the transmembranous domain of a receptor molecule at least partially determines its presence in ceramide-enriched membrane platforms.[44] Therefore, binding of a ligand to a receptor may trigger trapping of the receptor in ceramide-enriched membrane domains through conformational changes within the transmembranous domain, which then energetically favors the presence of this receptor in ceramide-enriched membrane platforms (Fig. 3).

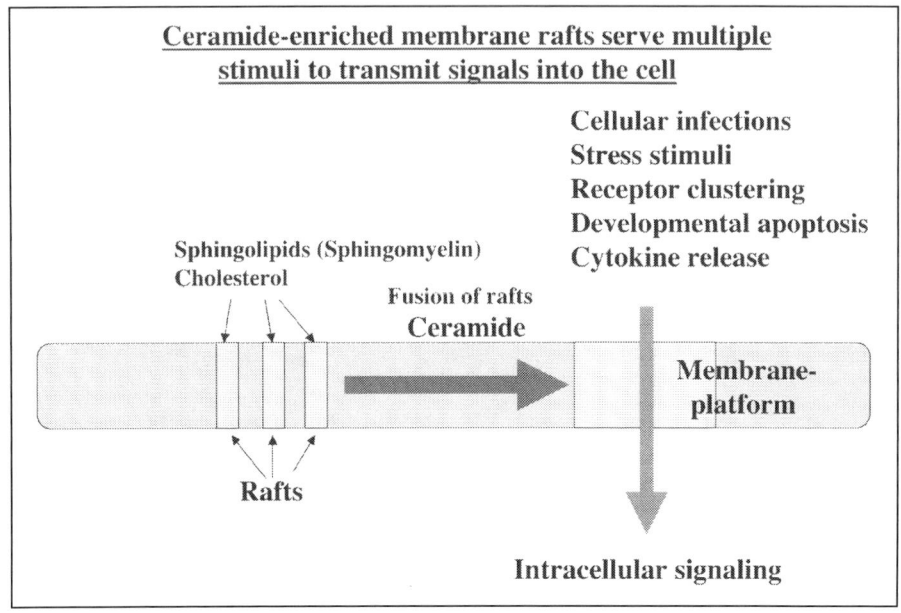

Figure 4. Ceramide-enriched membrane platforms mediate cellular stimulation by various stimuli. The generation of ceramide-enriched membrane platforms by the activity of the acid sphingomyelinase occurs upon treatment of cells with various stimuli including radiation, heat, UV-light, bacteria, viruses or during developmental death of oocytes. Although not formally proven for all of those stimuli, we hypothesize that ceramide-enriched membrane platforms are central for the transmission of the signal into the cell.

Ceramide-Enriched Membrane Platforms as General Signaling Motif

Since the formation of ceramide-enriched membrane platforms is not restricted to a single, specific signaling event, but rather acts as a universal amplifier, these structures are able to support signaling via various receptors and stimuli. Thus, a very similar concept has been shown to apply for signaling via the CD40 or TNF receptors that stimulate the acid sphingomyelinase and cluster in ceramide-enriched membrane platforms (refs. 14,31 and unpublished observations). Clustering facilitates these molecules to reach the critical density required for the transmission of the signal into the cell. Furthermore, ceramide-enriched membrane platforms are formed upon radiation,[46,47] UVA light,[48] heat shock,[49] treatment with some chemotherapeutics[50] and infections with pathogenic bacteria,[15,51-53] viruses[54] and parasites.[55] Although the details of signal initiation by these stimuli are unknown, these data suggest ceramide-enriched membrane platforms as structures performing a general function in signal transduction (Fig. 4).

Summary

Many recent data indicate the requirement of rafts and ceramide-enriched membrane domains for the induction of apoptosis upon CD95 stimulation. Ceramide-enriched membrane domains serve to cluster the receptor to reach the critical density that is required for transmission of the signal into the cell. Furthermore, they promote the assembly of the intra-cellular signaling machinery that transfers the signal from the receptor into the cell. Finally, ceramide-enriched membrane platforms might also serve to exclude and silence anti-apoptotic receptors and/or signaling molecules. The function of ceramide-enriched membrane domains is not restricted to CD95, but seems to apply to many other receptor molecules and nonreceptor stimuli, e.g., radiation, heat shock, UV-light or with infections with pathogens.

Acknowledgements

The studies were supported by DFG Gu 335/13-1 to E.G.

References

1. Singer SJ, Nicolson GL. The fluid mosaic model of the structure of cell membranes. Science 1972; 175:720-731.
2. Simons K, Ikonen E. Functional rafts in cell membranes. Nature 1997; 387:569-572.
3. Brown DA, London E. Functions of lipid rafts in biological membranes. Annu Rev Cell Dev Biol 1998; 14:111-167.
4. Andersen RG. The caveolae membrane system. Annu Rev Biochem 1998; 67:199-225.
5. Harder T, Simons K. Caveolae, DIGs, and the dynamics of sphingolipid-cholesterol microdomains. Curr Opin Cell Biol 1997; 9:534-542.
6. Heerklotz H. Triton promotes domain formation in lipid raft mixtures. Biophys J 2002; 83:2693-2701.
7. Mayor S, Maxfield FR. Insolubility and redistribution of GPI-anchored proteins at the cell surface after detergent treatment. Mol Biol Cell 1995; 6:929-944.
8. Schuchman EH, Suchi M, Takahashi T et al. Human acid sphingomyelinase. Isolation, nucleotide sequence and expression of the full-length and alternatively spliced cDNAs. J Biol Chem 1991; 266:8531-8539.
9. Schissel SL, Jiang X, Tweedie-Hardman J et al. Secretory sphingomyelinase, a product of the acid sphingomyelinase gene, can hydrolyze atherogenic lipoproteins at neutral pH. Implications for atherosclerotic lesion development. J Biol Chem 1998; 273:2738-2746.
10. Duan RD, Hertervig E, Nyberg L et al. Distribution of alkaline sphingomyelinase activity in human beings and animals. Tissue and species differences. Dig Dis Sci 1996; 41:1801-1806.
11. Grassmé H, Jekle A, Riehle A et al. CD95 signaling via ceramide-rich membrane rafts. J Biol Chem 2001; 276:20589-20596.
12. Grassmé H, Schwarz H, Gulbins E. Surface ceramide mediates CD95 clustering. Biochem Biophys Res Commun 2001; 284:1016-1030.
13. Cremesti A, Paris F, Grassmé H et al. Ceramide enables Fas to cap and kill. J Biol Chem 2001; 276:23954-23961.
14. Grassmé H, Jendrossek V, Riehle A et al. Ceramide-rich membrane rafts mediate CD40 clustering. J Immunol 2001; 168:298-307.
15. Grassmé H, Jendrossek V, Riehle A et al. Host defense against P. aeruginosa requires ceramide-rich membrane rafts. Nat Med 2003; 9:322-330.
16. Qiu H, Edmunds T, Baker-Malcolm J et al. Activation of human acid sphingomyelinase through modification or deletion of C-terminal cysteine. J Biol Chem 2003; 278:32744-32752.
17. Kolzer M, Arenz C, Ferlinz K et al. Phosphatidylinositol-3,5-Bisphosphate is a potent and selective inhibitor of acid sphingomyelinase. Biol Chem 2003; 384:1293-1298.
18. Emmelot P, Van Hoeven RP. Phospholipid unsaturation and plasma membrane organization. Chem Phys Lipids 1975; 14:236-246.
19. Kolesnick RN, Goni FM, Alonso A. Compartmentalization of ceramide signaling: Physical foundations and biological effects. J Cell Physiol 2000; 184:285-300.
20. Holopainen JM, Subramanian M, Kinnunen PK. Sphingomyelinase induces lipid microdomain formation in a fluid phosphatidycholine/sphingomyelin membrane. Biochemistry 1998; 37:17562-17570.
21. Nurminen TA, Holopainen JM, Zhao H et al. Observation of topical catalysis by sphingomyelinase coupled to microspheres. J Am Chem Soc 202; 124:12129-12134.
22. Huang HW, Goldberg EM, Zidovetzki R. Ceramides modulate protein kinase C activity and perturb the structure of phosphatidylcholine/ phosphatidylserine bilayers. Biophys J 1999; 77:1489-1497.

23. Veiga MP, Arrondon JL, Goni FM et al. Ceramides in phospholipid membranes: Effects on bilayer stability and transition to nonlamellar phases. Biophys J 1999; 76:342-350.

24. ten Grotenhuis E, Demel RA, Ponec M et al. Phase behavior of stratum corneum lipids in mixed Langmuir-Blodgett monolayers. Biophys J 1996; 71:1389-1399.

25. Horinouchi K, Erlich S, Perl DP et al. Acid sphingomyelinase deficient mice: A model of types A and B Niemann-Pick disease. Nat Genet 1995; 10:288-293.

26. Callahan JW, Jones CS, Davidson DJ et al. The active site of lysosomal sphingomyelinase: Evidence for the involvement of hydrophobic and ionic groups. J Neurosci Res 1983; 10:151-163.

27. Grassmé H, Cremesti A, Kolesnick R et al. Ceramide-mediated clustering is required for CD95-DISC formation. Oncogene 2003; 22:5457-5470.

28. Fanzo JC, Lynch MP, Phee H et al. CD95 rapidly clusters in cells of diverse origins. Cancer Biology and Therapy 2003; 2:392-395.

29. Kirschnek S, Paris F, Weller M et al. CD95-mediated apoptosis in vivo involves acid sphingomyelinase. J Biol Chem 2002; 275:27316-27323.

30. Paris F, Grassmé H, Cremesti A et al. Natural ceramide reverses Fas resistance of acid sphingomyelinase (-/-) hepatocytes. J Biol Chem 2000; 276:8297-8305.

31. Garcia-Ruiz C, Colell A, Mari M et al. Defective TNF-alpha-mediated hepatocellular apoptosis and liver damage in acidic sphingomyelinase knockout mice. J Clin Invest 2003; 111:197-208.

32. Schütze S, Potthoff K, Machleidt T et al. TNF activates NF-kappa B by phosphatidylcholine-specific phospholipase C-induced "acidic" sphingomyelin breakdown. Cell 1992; 71:765-776.

33. Seino KI, Kayagaki N, Takeda K et al. Contribution of Fas ligand to T cell-mediated hepatic injury in mice. Gastroenterology 1997; 113:1315-1322.

34. Wang ZQ, Dudhane A, Orlikowski T et al. CD4 engagement induces Fas antigen-dependent apoptosis of T cells in vivo. Eur J Immunol 1994; 24:1549-1552.

35. Bock J, Szabò I, Gamper N et al. Ceramide inhibits the potassium channel Kv1.3 by the formation of membrane platforms. Biochem Biophys Res Comm 2003; 305:890-897.

36. Hueber AO, Bernard AM, Herincs Z et al. An essential role for membrane rafts in the initiation of Fas/CD95-triggered cell death in mouse thymocytes. EMBO Rep 2002; 3:190-196.

37. Garofalo T, Misasi R, Mattei V et al. Association of the death-inducing signaling complex with microdomains after triggering through CD95/Fas. Evidence for caspase-8-ganglioside interaction in T cells. J Biol Chem 2003; 278:8309-8315.

38. Scheel-Toellner D, Wang K, Singh R et al. The death-inducing signalling complex is recruited to lipid rafts in Fas-induced apoptosis. Biochem Biophys Res Commun 2002; 297:876-879.

39. Delmas D, Rebe C, Lacour S et al. Resveratrol-induced apoptosis is associated with Fas redistribution in the rafts and the formation of a death-inducing signaling complex in colon cancer cells. J Biol Chem 2003; 278:41482-41490.

40. Gulbins E, Szabo I, Baltzer K et al. Ceramide-induced inhibition of T lymphocyte voltage-gated potassium channel is mediated by tyrosine kinases. Proc Natl Acad Sci USA 1997; 94:7661-7666.

41. Bock J, Szabò I, Jekle A et al. Actinomycin D-induced apoptosis involves the potassium channel Kv1.3. Biochem Biophys Res Comm 2002; 295:526-531.

42. Bini L, Pacini S, Liberatori S et al. Extensive temporally regulated reorganization of the lipid raft proteome following T-cell antigen receptor triggering. Biochem J 2003; 369:301-309.

43. Megha, London E. Ceramide selectively displaces cholesterol from ordered lipid domains (Rafts): Implications for raft structure and function. J Biol Chem 2003 Dec 29 [Epub ahead of print].

44. Bock J, Gulbins E. The transmembranous domain of CD40 determines CD40 partitioning into lipid rafts. FEBS-Letters 2002; 534:169-174.

45. Janes PW, Ley SC, Magee AI. Aggregation of lipid rafts accompanies signaling via the T cell antigen receptor. J Cell Biol 1999; 147:447-461.

46. Santana P, Pena LA, Haimovitz-Friedman A et al. Acid sphingomyelinase-deficient human lymphoblasts and mice are defective in radiation-induced apoptosis. Cell 1996; 86:189-199.

47. Garcia-Barros M, Paris F, Cordon-Cardo C et al. Tumor response to radiotherapy regulated by endothelial cell apoptosis. Science 2003; 300:1155-1159.

48. Zhang Y, Mattjus P, Schmid PC et al. Involvement of the acid sphingomyelinase pathway in UVA-induced apoptosis. J Biol Chem 2001; 276:11775-11782.

49. Chung HS, Park SR, Choi EK et al. Role of sphingomyelin-MAPKs pathway in heat-induced apoptosis. Exp Mol Med 2003; 35:181-188.
50. Morita Y, Perez GI, Paris F et al. Oocyte apoptosis is suppressed by disruption of the acid sphingomyelinase gene or by sphingosine-1- phosphate therapy. Nat Med 2000; 6:1109-1114.
51. Grassmé H, Gulbins E, Brenner B et al. Acidic sphingomyelinase mediates entry of N. gonorrhoeae into nonphagocytic cells. Cell 1997; 91:605-615.
52. Hauck CR, Grassmé H, Bock J et al. Acid sphingomyelinase is involved in CEACAM receptor-mediated phagocytosis of N. gonorrhoeae. FEBS-Letters 2000; 478:260-266.
53. Esen M, Schreiner B, Jendrossek V et al. Mechanisms of Staphylococcus aureus induced apoptosis of human endothelial cells. Apoptosis 2001; 6:441-445.
54. Jan JT, Chatterjee S, Griffin DE. Sindbis virus entry into cells triggers apoptosis by activating sphingomyelinase, leading to the release of ceramide. J Virology 2000; 74:6425-6432.
55. Hanada K, Palacpac NM, Magistrado PA et al. Plasmodium falciparum phospholipase C hydrolyzing sphingomyelin and lysocholinephospholipids is a possible target for malaria chemotherapy. J Exp Med 2002; 195:23-34.

CHAPTER 4

Regulation of Fas Signaling by FLIP Proteins

Margot Thome

Abstract

Fas is a member of the tumor necrosis factor receptor family that can induce apoptosis by the recruitment and activation of caspase-8 (formerly called FLICE, MACH or MCH-5). Recently, caspase-8/FLICE inhibitory proteins (FLIPs) have been identified as proteins that counteract the caspase-8-dependent apoptosis-promoting activity of Fas and other death receptors. Viral and cellular FLIPs, which share structural similarity with caspase-8, are recruited to the death receptors upon ligand binding and inhibit caspase-8 activation. Viral FLIP family members are present in several lymphotropic herpesviruses and in a human poxvirus, and expression of viral FLIP proteins is thought to prevent or delay the elimination of virus-infected cells by cytotoxic T cells. Cellular FLIP has a similar anti-apoptotic function, but genetic studies have revealed additional, previously unanticipated roles in T-cell proliferation and heart development. Moreover, abnormal expression of cellular FLIP may play a role in autoimmune diseases, in tumor development and in cardiovascular disorders.

Abbreviations

AICD, activation induced cell death; ALPS, autoimmune lymphoproliferative syndrome; CARD, caspase recruitment domain; DD, death domain; DED, death effector domain; DISC, death-inducing signaling complex; FLIP, FLICE/caspase-8 inhibitory protein; MAPK, mitogen-activated protein kinases; NF-κB, nuclear factor kappa-B; RIP, receptor interacting protein; TNFR, tumor necrosis factor receptor; TRAF, TNF receptor-associated factor; TRAIL, TNF-related apoptosis-inducing ligand.

Introduction

The molecular mechanism of Fas signaling has been studied extensively.[1] Fas has a cytoplasmic death domain (DD), and binding of the membrane-bound, trimeric form of FasL on an effector cell to the receptor on the target cell results in the recruitment of the adapter molecule FADD (Fas-associated protein with death-domain) to the receptor via a DD/DD interaction.[2,3] FADD also contains an N-terminal death effector domain (DED), which in turn recruits the DED-containing caspases-8 and -10,[4-8] thereby forming a receptor-bound complex called the death-inducing signaling complex (DISC).[1] It is thought that this process brings two or more caspase-8 (or -10) molecules into close proximity at the receptor level, and allows them to proteolytically activate each other.[9-12] Active caspase molecules that lack the N-terminal DEDs are then released into the cytoplasm and initiate apoptosis by subsequent

Fas Signaling, edited by Harald Wajant. ©2006 Landes Bioscience and Springer Science+Business Media.

cleavage and activation of downstream effector caspases (caspases-3, -6, -7) or of the pro-apoptotic Bcl-2 family member, Bid.[1]

What Are the Structural Features of FLIP Family Proteins That Contribute to Their Anti-Apoptotic Function?

FLIP proteins were initially identified by means of a bioinformatic screen for novel proteins with a DED. All FLIP family members are characterized by the presence of two N-terminal DEDs, which are followed by a C-terminal portion that varies considerably between family members (Fig. 1). Viral FLIP (v-FLIP) proteins were first detected in γ2-herpesviruses such as equine herpesvirus-2 (EHV-2), herpesvirus saimiri (HVS), Kaposi sarcoma-associated herpesvirus (human herpesvirus-8, HHV-8)[13-15] and, subsequently, in rhesus rhadinovirus (RRV).[16] The human poxvirus molluscum contagiosum virus (MCV) contains two quite different v-FLIP variants with unusually long C-terminal extensions of unknown function.[13-15] Cellular FLIP (c-FLIP, also called CASH, Casper, CLARP, FLAME, I-FLICE, MRIT or usurpin), was subsequently identified as a close homologue of the v-FLIP molecules.[17-24] Several splice isoforms of c-FLIP have been described,[17,24] two of which have been shown to be expressed at the protein level. The short FLIP (FLIP$_S$) isoform of 26 kD is similar to the v-FLIPs of the γ-herpesvirus-type in its N-terminal structure, but contains a unique C-terminal extension of 25 amino acids with unknown function. The long FLIP (FLIP$_L$) isoform of 55 kD has a longer C-terminal extension that includes a region with homology to caspases. The overall structure of FLIP$_L$ is thus similar to caspase-8 and -10, but FLIP$_L$ lacks catalytic activity because several of the amino acids that are critically required for catalytic activity[25] have not been conserved. However, FLIP$_L$ contains an Asp residue (Asp 341) between the p20 and p10-like domains of the caspase-homologous region,[19] that is in a position which is well conserved in caspases-8 and -10 and can be cleaved in the context of Fas-signaling (Fig. 2A).[17,26] Together, these structural features of FLIP molecules predicted that they could interfere with caspase-8 and caspase-10-dependent pathways, by impeding caspase-8 or -10 recruitment to the receptor and/or by directly binding and inhibiting these caspases.

What Is the Molecular Function of Viral and Cellular FLIP?

Owing to their structural homology with the N-terminus of caspase-8 and -10, but their highly variable C-terminal extensions, FLIP$_S$ and FLIP$_L$ proteins can interfere with the receptor-mediated activation of these caspases by different mechanisms (Fig. 2, A-C). v- and c-FLIP$_{S/L}$ bind to the DED-containing amino-terminus of FADD, caspase-8 or caspase-10[27] and are thereby recruited to the activated Fas receptor.[15,17,19,26] Viral and cellular FLIP$_S$ act as dominant negative inhibitors of caspase-8 by prevention of the processing and release of active caspase-8 from the receptor[15,28] (Fig. 2B). FLIP$_L$, on the other hand, can bind to both the DEDs and the caspase domain of caspase-8.[17,19,22,24] Moreover, it has a somewhat different effect than FLIP$_S$ on the receptor-associated signaling complex. Indeed, in the presence of FLIP$_L$, both, caspase-8 and FLIP$_L$ are partially processed into a C-terminal p10 fragment and an N-terminal p43 fragment that stay bound to the receptor complex[26,28,29] (Fig. 2C).

Strikingly, FLIP$_L$ is a much more potent inhibitor of death receptor-induced cell death than FLIP$_S$ at similar expression levels[17] (Fig. 3). While FLIP$_S$ partially protects from FasL- and TRAIL-induced cytotoxicity, FLIP$_L$ provides a considerably stronger or complete protection against FasL- and TRAIL-induced cell death, respectively. What are the molecular reasons for the increased cell viability in the presence of FLIP$_L$?

One possible explanation comes from the observation that, in the presence of FLIP$_L$, additional signaling molecules such as TNF receptor-associated factor-1 and -2 (TRAF1 and -2), and Raf-1 are also recruited/stabilized within the Fas-DISC.[30] These proteins may

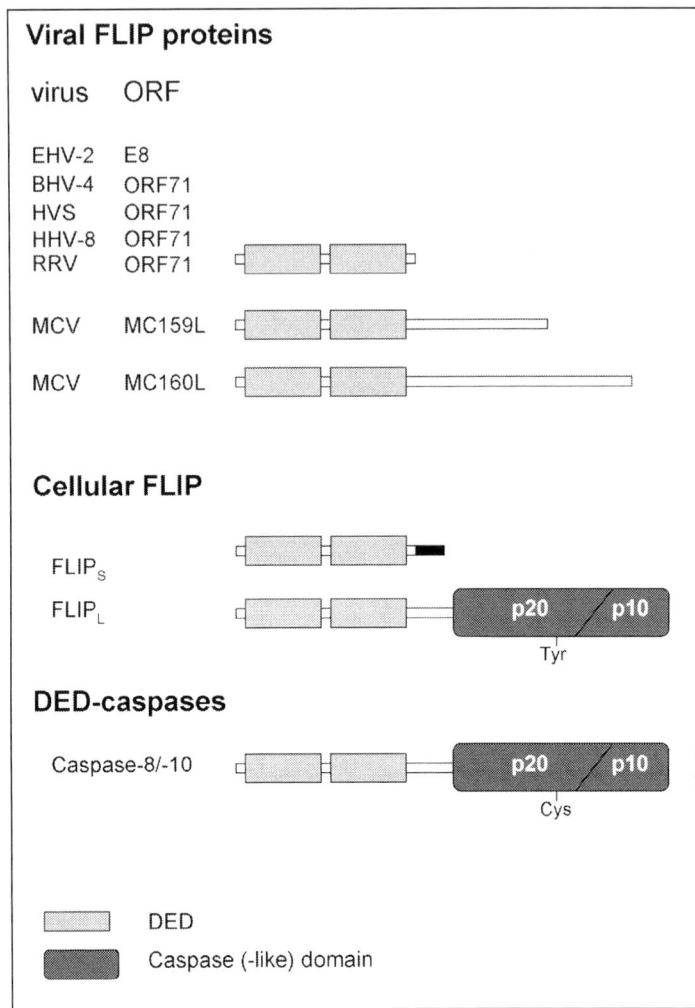

Figure 1. Overview of the molecular structure of FLIP proteins. The v-FLIP proteins of EHV-2, BHV-4, HHV-8, HVS and RRV consist essentially of two repeats of a death effector domain (DED). Two additional v-FLIP variants encoded by ORFs found in the human molluscipox virus (MC159L/160L) have C-terminal extensions of unknown function. The short form of c-FLIP (FLIP$_S$) contains two DEDs and a unique C-terminal extension of 25 amino acids of unknown function. The long splice variant of c-FLIP (FLIP$_L$) contains a C-terminal caspase-like domain that is catalytically inactive because it lacks the critical active site Cys residue and other conserved caspase-typical amino acids. The overall structure of FLIP$_L$ is thus similar to caspase-8 and -10.

link FLIP$_L$ to the activation of the nuclear factor-kappa B (NF-κB) and extracellular signal-regulated kinase (ERK)-dependent transcriptional pathways and thus lead to increased proliferation and/or survival of FLIP$_L$-expressing Fas-stimulated cells (Fig. 4A). Since FLIP expression is induced by NF-κB itself, such a mechanism could serve as a feedback amplification loop by further increasing FLIP expression levels.[31]

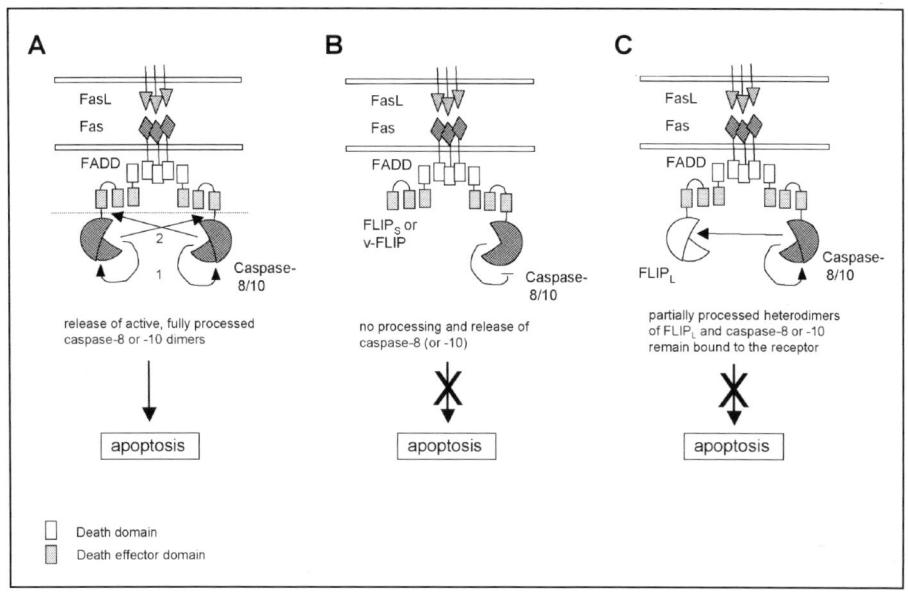

Figure 2. Fas signaling in the absence or presence of FLIP. Binding of the membrane-bound trimeric Fas ligand (FasL) to the Fas receptor leads to recruitment of the adapter molecule FADD via a homophilic interaction between the death domain (DD) of the receptor and FADD. The death effector domain (DED) of FADD serves to recruit caspase-8 via DED/DED interactions. A) In the absence of FLIP, caspase-8 is recruited and activated by autocatalytic cleavage (1) in the caspase-domain between the p20 and the p10 portions and (2) at a site following the DEDs and preceding the caspase domain. As a result, the DED-containing amino-terminal fragment of caspase-8 stays transiently at the DISC, while the active caspase-8 protease-dimer is released into the cytoplasm to initiate the apoptotic cascade. B) The viral FLIPs and the short form of cellular FLIP (FLIP$_S$) have two DEDs and bind to FADD and/or caspase-8. Thereby, both v-FLIP and FLIP$_S$ inhibit the processing of caspase-8 at the receptor level and protect the cells from apoptosis. C) Cellular long FLIP (FLIP$_L$) has two DEDs and a caspase-like domain that lacks catalytic activity due to the absence of cysteine residue. In the presence of FLIP$_L$, both caspase-8 and FLIP are recruited and partially processed at the receptor level. The partially processed caspase-8 and FLIP proteins stay bound to the receptor and no fully processed caspase-8 can be released into the cytosol. The exact stochiometry of the receptor-induced signaling complexes is unknown.

Another hypothesis that may explain the differential behavior of FLIP$_L$ versus FLIP$_S$ is linked to the observation that in the presence of FLIP$_L$, but not of FLIP$_S$, both caspase-8 and FLIP$_L$ undergo partial processing into a C-terminal p10 and an N-terminal p43 fragment within the DISC. Surprisingly, this Fas-associated caspase-8/FLIP$_L$ complex has substantial enzymatic activity.[29,32] It is conceivable that the partially processed caspase-8/FLIP$_L$ heterodimer has a proteolytic activity with altered substrate specificity or membrane-restricted action, which could increase cell survival by cleavage-induced inactivation of cell death-promoting proteins. This hypothesis is interesting in light of the recent observation that death receptor-induced cell death can be mediated by both, caspase-dependent (apoptotic) and caspase-independent (necrotic) mechanisms that depend on the catalytic activity of the DD-containing receptor-interacting protein (RIP) kinase.[33,34] It thus appears that if FLIP$_S$ and FLIP$_L$ both inhibit apoptosis to similar extents, then the increased protection against Fas-induced cell death conferred by FLIP$_L$ could be the result of an anti-necrotic function of FLIP$_L$ (Fig. 4B). Consistent with this hypothesis, FLIP$_L$ was found to favor Fas- induced cleavage of the pro-necrotic

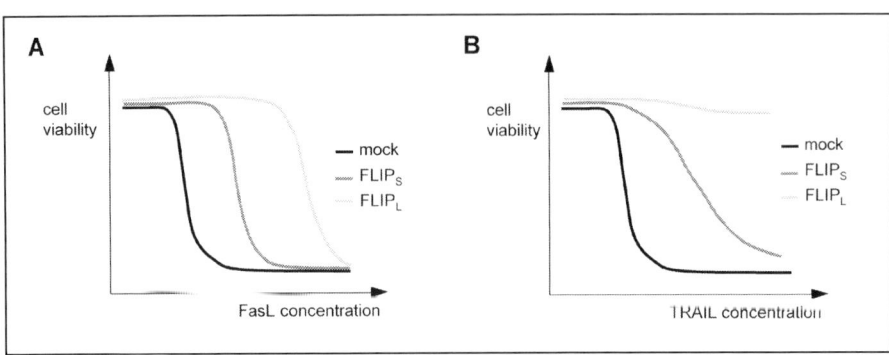

Figure 3. FLIP$_L$ protects more potently against death receptor-induced cell death than FLIP$_S$. Schematically summarized are the results of FLIP$_S$- and FLIP$_L$-provided protection of Jurkat T cells against FasL- or TRAIL-induced apoptosis. Jurkat cells stably transfected with the short or long form of FLIP or with control vector (mock) were treated with crosslinked FasL (A) or TRAIL (B), and cell death quantified by staining of surviving cells with a metabolic dye coloring viable, metabolizing cells. At similar expression levels, FLIP$_L$ protected cells more efficiently than FLIP$_S$ against FasL- or TRAIL-induced cell death.[17]

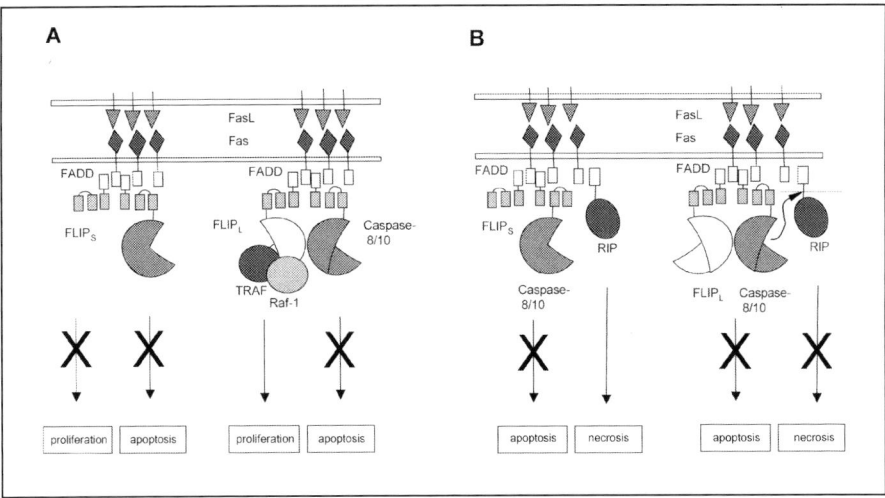

Figure 4. Hypothetical model for the increased cell viability of FLIP$_L$-expressing cells. A) In the presence of FLIP$_L$, but not FLIP$_S$, additional signaling molecules such as TRAF1, TRAF2 and raf-1 are recruited to the receptor complex and may contribute to protection from cell death by induction of proliferation- or differentiation-inducing transcriptional pathways. B) The catalytically active caspase-8/FLIP$_L$ heterodimer may protect the cells from Fas-induced necrosis by the proteolytic degradation of membrane-proximal pro-necrotic molecules such as RIP. Both mechanisms may contribute to the increased viability of FLIP$_L$- versus FLIP$_S$ -expressing cells after death receptor-stimulation.

protein RIP.[29] Interestingly, the v-FLIPs MC159 (of the molluscipox virus) and E8 (of the equine herpes virus-2) have been shown to suppress both apoptosis and necrosis induced by TNF-RI.[33,34] Obviously, more than one of the above outlined mechanisms may contribute to the increased cell death-preventing activity of FLIP$_L$ versus FLIP$_S$.

Further studies will be necessary to dissect the specific molecular functions of $FLIP_S$ and $FLIP_L$, for example through the generation of mice that express only $FLIP_S$ and not $FLIP_L$, and by biochemical identification and functional characterization of $FLIP_L$- (but not $FLIP_S$-) associated proteins.

What Is the Biological Function of the FLIP Proteins?

The Fas/FasL system has an important role in the immune system where FasL-expressing cytotoxic T cells mediate the killing of Fas-expressing virally infected or tumor target cells.[35,36] Moreover, Fas-mediated cell death also plays a role in downsizing the peripheral B- and T-cell pool after activation and proliferation, in order to prevent accumulation of T- and B-cells, production of autoantibodies and development of autoimmune disease. This is evident from studies of Fas or FasL deficient (lpr and gld) mice or of humans with autoimmune lymphoproliferative syndrome (ALPS), who have mutations in Fas, FasL or other genes in the Fas pathway. Both the human and mouse mutations produce an accumulation of peripheral B- and T cells and autoimmune symptoms (reviewed in ref. 37). It was thus expected that c-FLIP-transgenic or -deficient mice would interfere with these biological functions by disturbing cytotoxic T cell-dependent immune responses or by interfering with the efficient elimination of activated B- and T-cells. c-FLIP was also expected to have additional functions, since FLIP expression is abundant not only in the lymphoid system, but also in the heart and skeletal muscle and in the kidney.[17,24] FLIP-deficient mice have been generated and fibroblast cell lines derived from these animals showed increased sensitivity to FasL- and TNF-induced apoptosis, consistent with a protective role for FLIP in death receptor-mediated apoptosis.[38] Moreover, the knock-out mice revealed an important function for FLIP in embryonic development, since FLIP-deficient animals showed signs of cardiac failure and impaired heart development, and did not survive past day 10.5 of embryogenesis (Table 1). This observation correlates with the fact that c-FLIP is abundantly expressed in the adult human and murine heart.[24] Since both, FADD and caspase-8, but not Fas or TNFR-1 deficient mice, show symptoms of impaired heart development,[39-42] these findings suggest that a signaling complex involving FADD, caspase-8, FLIP and possibly a death domain containing receptor other than Fas or TNFR-1 may be involved in heart development.

The embryonic lethality of the FLIP-deficient mice has made it difficult to assess the function of FLIP in the lymphocyte compartment, but three independent laboratories have taken an alternative approach to study FLIP function by overexpression of FLIP in the lymphocytes of mice. The studies by Lens et al and by Tai et al analyzed FLIP-transgenic mice that specifically overexpress $FLIP_L$ in the T-cell compartment.[43,44] In both studies, peripheral T cells of the mice were protected against Fas-induced apoptosis in vitro, but, surprisingly, there was no lpr/gld-like accumulation of activated T cells in vivo. Differing results were obtained, however, with respect to the effect of FLIP on T-cell activation and proliferation. In the study by Lens et al, $FLIP_L$-transgenic T cells responded with increased proliferation to sub-optimal doses of TCR-triggering anti-CD3 antibodies, but showed reduced proliferation and increased apoptotic features at higher doses of anti-CD3. In contrast, in the $FLIP_L$ T-cell transgenic mice generated by Tai et al[44] the peripheral T cells showed clearly reduced proliferation and IL-2 production at all stimulating anti-CD3 concentrations used. The reason for the differing results of the two studies is unclear, it may be due to differences in the anti-CD3 stimulation protocol, differences in the expression levels of the transgene with respect to the endogenous protein, or differences in the relative expression of $FLIP_L$ and $FLIP_S$ from the transgenic construct.

In order to address the biological function of FLIP in B-cells, Van Parijs et al have taken a retroviral approach to overexpress c-$FLIP_L$ in isolated bone marrow cells that were used to reconstitute hematopoietic cells of lethally irradiated recipient mice.[45] In this experimental

Table 1. *Genetic evidences for the biological functions of FADD, caspase-8, caspase-10 and FLIP*

Genetic Alteration	Heart Development	Lympho-Proliferative Symptoms	T-Cell Activation	Refs.
FADD[-/-]	impaired	n.t.	n.t.	39
FADD-DN rag-1[-/-]	normal	no	impaired	40
FADD-DN (T cell-transgenic)	normal	no	impaired	46,47,48, 51,76
Caspase-8[-/-]	impaired	n.t.	n.t.	41
Caspase-8[-/-] (T cells)	normal	no	impaired	49
Caspase-8 defect (human)	normal	yes	impaired	50
Caspase-10 defect (human)	normal	yes	n.t.	77
FLIP[-/-]	impaired	n.t.	n.t.	38
FLIP (T-cell transgenic)	normal	no	Increased/impaired at suboptimal/high anti-CD3 doses,	43
FLIP (T-cell trangenic)	normal	no	Impaired at all anti-CD3 concentrations	44
FLIP overexpression (bone marrow)	normal	Accumulation of mainly B-cells	Unaltered	45

Summarized are the phenotypic features of human and mouse genetic alterations in the expression of these signaling molecules. For those mutants that are embryonically lethal, lymphocyte functions could not be assessed. Abbreviations used: dominant negative (DN), not tested (n.t.).

model, $FLIP_L$-expressing B- and T-cells were resistant to Fas-mediated cell death. Of note, older mice accumulated B- and T-cells and developed autoimmunity.[45] These findings suggested that downregulation of c-FLIP may be necessary to prevent accumulation of lymphocytes and thus to maintain self-tolerance by elimination of autoreactive cells.

How do these findings compare to genetic studies of other molecules in the Fas signaling pathway? FADD-deficient mice show embryonic lethality, but reconstitution of RAG-2[-/-] mice with FADD[-/-] ES cells and the generation of mice expressing a dominant negative form of FADD (DN-FADD) have allowed the function of FADD in lymphocyte function to be addressed. Surprisingly, both types of mice were immunodeficient because of a defect in T- cell activation, and did not show lymphoproliferative symptoms.[40,46-48] Moreover, immunodeficiency with impaired T-cell proliferative responses have recently been reported for T-cell specific caspase-8 knock-out mice[49] and for two human patients carrying an inactivating caspase-8 mutation.[50] While the human patients also showed characteristics of ALPS, no lymphoproliferative syndrome was detected in the T-cell specific caspase-8 mutant mice. Together, the results of these genetic studies suggest that (1) FADD, caspase-8 and potentially FLIP play a role in the regulation of T-cell activation and proliferation, and (2) that the accumulation of T cells in lpr/gld mice or ALPS patients may not exclusively result from a T-cell intrinsic defect in the Fas-FADD-caspase-8 pathway, but rather depend on the inhibition of the Fas pathway in multiple cell types of the immune system. Additional studies, such as the generation of conditional knock-out mice that specifically deplete FLIP expression in T cells will hopefully provide insight into the exact role of c-FLIP in T-cell proliferation. It also remains to be established whether the observed effects of FADD, caspase-8 and FLIP are due to a direct effect of these proteins on T-cell activation or TNFR-family signaling pathways or whether developmental defects of immature T-cells may account for the altered proliferative response of mature T cells. Indeed, in mice expressing a dominant negative form of FADD, thymocyte development was impaired at the CD25⁺CD44⁻ (DN3) stage,[46,51,52] also called the preTCR checkpoint stage because the cells at this stage depend on intact preTCR-signaling for the induction of thymocyte proliferation. Consequently, the mice had reduced numbers of thymocytes and peripheral T cells, suggesting that FADD is also critically involved in the development of functional mature T cells.

Finally, several biochemical studies have addressed the potential biological function of FLIP by monitoring its expression level in various functional subsets of immune cells. $FLIP_L$ was found to be highly expressed in freshly activated T-cells, but was down-regulated upon prolonged stimulation in the presence of IL-2, suggesting that decreased FLIP expression levels may sensitize activated T cells to Fas-induced apoptosis and thus to the termination of the immune response by activation-induced cell death.[17,45,53] On the other hand, increased expression levels of $c-FLIP_S$ have been found in memory T-cell models in vitro, suggesting that upregulation of FLIP could contribute to the increased FasL-resistance of memory T cells.[54,55]

FLIP-expression may also regulate the Fas sensitivity of monocytes and monocyte-derived cells. While monocytes are sensitive to Fas-induced apoptosis, monocyte-derived macrophages and dendritic cells (DCs) acquire Fas-resistance that correlates with an upregulation of FLIP expression.[56-58] DCs that present antigen to T cells could thus be protected from Fas-induced apoptosis by FLIP. Moreover, increased FLIP expression in macrophages may contribute to the development of certain inflammatory diseases that are characterized by accumulation of activated macrophages in the diseased tissues. Finally, Fas engagement induces the maturation of constitutively FLIP-expressing dendritic cells, suggesting that FLIP may not only protect the cells from apoptosis but also participate in a Fas-dependent differentiation process.[59]

Is Altered FLIP Expression Linked to Pathologies?

Altered c-FLIP expression levels have been observed in several pathologies, especially in autoimmune and inflammatory diseases, but also in cardiovascular diseases and in certain forms of cancer[27,60] (Table 2). In the context of autoimmune diseases, pathological upregulation of FLIP levels may contribute to a diminished elimination of pathogenic, autoreactive lymphocytes. This has been suggested to be the case in multiple sclerosis,[61,62] a neuro-inflammatory disease in which autoreactive T cells may respond to myelin self antigens.[63] In rheumatoid arthritis, characterized by chronic inflammation of the synovial joints, increased FLIP expression levels have been observed in synovial macrophages that are thought to contribute to the inflammatory symptoms by the secretion of the pro-inflammatory cytokines TNF and IL-1.[64] Finally, in a form of autoimmune hyperthyroidism called Grave's disease,[65] increased FLIP levels in the thyrocytes themselves may be linked to the abnormal thyrocyte proliferation in this disease.

Abnormal downregulation of FLIP expression levels, on the other hand, has been observed in certain cardiovascular pathologies. FLIP was found to be specifically downregulated in an animal model of ischemia/reperfusion injury-induced cardiac infarcts and in apoptotic cardiomyocyte grafts, suggesting that downregulation of FLIP may sensitize cardiac myocytes to apoptotic death.[24,66] Moreover, FLIP downmodulation was observed in human myocardium samples from explanted cardiomyopathic hearts of patients with heart failure.[67]

Downregulation of FLIP expression levels could also play an important role under conditions that lead to pathological vascular tissue destruction.[68,69] Normal vascular endothelial cells express both Fas and FasL but are resistant to Fas-mediated apoptosis potentially because

Table 2. Examples of pathologies associated with disregulated FLIP expression.
Summarized are examples of pathologies in which c-FLIP expression levels have been shown to be upregulated (+) or downregulated (-)

Pathology	FLIP Level	Cell Type	Refs.
multiple sclerosis	+	T cells	61,62
Grave's disease	+	thyrocytes	65
rheumatoid arthritis	+	synovial macrophages	78
artherosclerosis	-	smooth muscle cell	68
atherogenesis	-	vascular endothelial cells	69
heart failure	-	cardiomyocytes	67
colon adenocarcinoma	+	carcinoma cells	74
melanoma	+	melanoma cells	17
B-cell lymphoma	+	Hodgkin cells Reed-Sternberg cell	75

they express c-FLIP.[69] Oxidized low density lipoproteins induce downregulation of FLIP and thereby sensitize cells to Fas- mediated apoptosis.[69] Finally, c-FLIP was shown to be expressed in endothelial cells, macrophages, and smooth muscle cells of human coronary arteries, but was down-regulated in smooth muscle cells within atherosclerotic plaques.[68] Thus, therapeutic regulation of FLIP expression levels could potentially be useful for the treatment of cardiovascular diseases such as atherogenesis and atherosclerosis.

Finally, it is possible that FLIP proteins function as tumor progression factors, since elevated FLIP expression in tumor cells correlated with tumor escape from T-cell and NK-cell immunity and enhanced tumor progression in vivo.[70-73] Moreover, increased c-FLIP expression levels have been reported for colon adenocarcinomas,[74] melanomas[17] and certain forms of B-cell lymphomas,[75] but a causative link between increased FLIP expression and the progression of naturally occurring tumors remains to be established.

Conclusions and Open Issues

The molecular mechanism of FLIP's antiapoptotic function in the context of death receptor signaling is relatively well understood, but several aspects of its molecular and physiological function remain to be resolved. Some viral FLIPs and the cellular $FLIP_L$ have C-terminal extensions whose exact molecular contribution to cell death protection is only partly understood. Moreover, it is unclear how FLIP contributes to the development and function of the heart and whether this implies an interaction of FLIP with a specific receptor/adapter system. In the immune system, FLIP may control the proliferation and/or survival of B-, T- and monocyte/macrophage lineage cells and thereby have an important role in the regulation of the immune response. Finally, a growing number of pathologies have been associated with disregulated FLIP expression. Monitoring and/or altering FLIP expression levels may thus be of therapeutic benefit in many circumstances.

Acknowledgments

The author would like to thank Helen Everett, Olivier Micheau and Mathias Thurau for helpful comments on the manuscript.

References

1. Peter ME, Krammer PH. The CD95(APO-1/Fas) DISC and beyond. Cell Death Differ 2003; 10:26-35.
2. Boldin MP, Varfolomeev EE, Pancer Z et al. A novel protein that interacts with the death domain of Fas/APO1 contains a sequence motif related to the death domain. J Biol Chem 1995; 270:7795-7798.
3. Chinnaiyan AM, O'Rourke K, Tewari M et al. FADD, a novel death domain-containing protein, interacts with the death domain of Fas and initiates apoptosis. Cell 1995; 81:505-512.
4. Boldin MP, Goncharov TM, Goltsev YV et al. Involvement of MACH, a novel MORT1/FADD-interacting protease, in Fas/APO-1- and TNF receptor-induced cell death. Cell 1996; 85:803-815.
5. Muzio M, Chinnaiyan AM, Kischkel FC et al. FLICE, a novel FADD-homologous ICE/CED-3-like protease, is recruited to the CD95 (Fas/APO-1) death—inducing signaling complex. Cell 1996; 85:817-827.
6. Kischkel FC, Lawrence DA, Tinel A et al. Death receptor recruitment of endogenous caspase-10 and apoptosis initiation in the absence of caspase-8. J Biol Chem 2001; 276:46639-46646.
7. Wang J, Chun HJ, Wong W et al. Caspase-10 is an initiator caspase in death receptor signaling. Proc Natl Acad Sci USA 2001; 98:13884-13888.
8. Sprick MR, Rieser E, Stahl H et al. Caspase-10 is recruited to and activated at the native TRAIL and CD95 death-inducing signalling complexes in a FADD-dependent manner but can not functionally substitute caspase-8. Embo J 2002; 21:4520-4530.
9. Medema JP, Scaffidi C, Kischkel FC et al. FLICE is activated by association with the CD95 death-inducing signaling complex (DISC). Embo J 1997; 16:2794-2804.
10. Yang X, Chang HY, Baltimore D. Autoproteolytic activation of pro-caspases by oligomerization. Mol Cell 1998; 1:319-325.
11. Martin DA, Siegel RM, Zheng L et al. Membrane oligomerization and cleavage activates the caspase-8 (FLICE/MACHalpha1) death signal. J Biol Chem 1998; 273:4345-4349.
12. Muzio M, Stockwell BR, Stennicke HR et al. An induced proximity model for caspase-8 activation. J Biol Chem 1998; 273:2926-2930.
13. Hu S, Vincenz C, Buller M et al. A novel family of viral death effector domain-containing molecules that inhibit both CD-95- and tumor necrosis factor receptor-1-induced apoptosis. J Biol Chem 1997; 272:9621-9624.
14. Bertin J, Armstrong RC, Ottilie S et al. Death effector domain-containing herpesvirus and poxvirus proteins inhibit both Fas- and TNFR1-induced apoptosis. Proc Natl Acad Sci USA 1997; 94:1172-1176.
15. Thome M, Schneider P, Hofmann K et al. Viral FLICE-inhibitory proteins (FLIPs) prevent apoptosis induced by death receptors. Nature 1997; 386:517-521.
16. Searles RP, Bergquam EP, Axthelm MK et al. Sequence and genomic analysis of a Rhesus macaque rhadinovirus with similarity to Kaposi's sarcoma-associated herpesvirus/human herpesvirus 8. J Virol 1999; 73:3040-3053.
17. Irmler M, Thome M, Hahne M et al. Inhibition of death receptor signals by cellular FLIP. Nature 1997; 388:190-195.
18. Shu HB, Halpin DR, Goeddel DV. Casper is a FADD- and caspase-related inducer of apoptosis. Immunity 1997; 6:751-763.
19. Srinivasula SM, Ahmad M, Ottilie S et al. FLAME-1, a novel FADD-like anti-apoptotic molecule that regulates Fas/TNFR1-induced apoptosis. J Biol Chem 1997; 272:18542-18545.
20. Inohara N, Koseki T, Hu Y et al. CLARP, a death effector domain-containing protein interacts with caspase-8 and regulates apoptosis. Proc Natl Acad Sci USA 1997; 94:10717-10722.
21. Goltsev YV, Kovalenko AV, Arnold E et al. CASH, a novel caspase homologue with death effector domains. J Biol Chem 1997; 272:19641-19644.
22. Han DK, Chaudhary PM, Wright ME et al. MRIT, a novel death-effector domain-containing protein, interacts with caspases and BclXL and initiates cell death. Proc Natl Acad Sci USA 1997; 94:11333-11338.
23. Hu S, Vincenz C, Ni J et al. I-FLICE, a novel inhibitor of tumor necrosis factor receptor-1- and CD- 95-induced apoptosis. J Biol Chem 1997; 272:17255-17257.

24. Rasper DM, Vaillancourt JP, Hadano S et al. Cell death attenuation by 'Usurpin', a mammalian DED-caspase homologue that precludes caspase-8 recruitment and activation by the CD-95 (Fas, APO-1) receptor complex. Cell Death Differ 1998; 5:271-288.

25. Cohen GM. Caspases: The executioners of apoptosis. Biochem J 1997; 326(Pt 1):1-16.

26. Scaffidi C, Schmitz I, Krammer PH et al. The role of c-FLIP in modulation of CD95-induced apoptosis. J Biol Chem 1999; 274:1541-1548.

27. Thome M, Tschopp J. Regulation of lymphocyte proliferation and death by FLIP. Nat Rev Immunol. 2001; 1:50-58.

28. Krueger A, Schmitz I, Baumann S et al. c-FLIP splice variants inhibit different steps of caspase-8 activation at the CD95 death-inducing signaling complex (DISC). J Biol Chem 2001; 5:5.

29. Micheau O, Thome M, Schneider P et al. The long form of FLIP is an activator of caspase-8 at the Fas death-inducing signaling complex. J Biol Chem. 2002; 277:45162-45171.

30. Kataoka T, Budd RC, Holler N et al. The caspase-8 inhibitor FLIP promotes activation of NF-kappaB and Erk signaling pathways. Curr Biol 2000; 10:640-648.

31. Micheau O, Lens S, Gaide O et al. NF-kB signals induce the expression of c-FLIP. Molecular and Cellular Biology 2001; 21:5299-5305.

32. Chang DW, Xing Z, Pan Y et al. c-FLIP(L) is a dual function regulator for caspase-8 activation and CD95-mediated apoptosis. Embo J 2002; 21:3704-3714.

33. Holler N, Zaru R, Micheau O et al. Fas triggers an alternative, caspase-8-independent cell death pathway using the kinase RIP as effector molecule. Nature Immunology 2000; 1:489-495.

34. Chan FK, Shisler J, Bixby JG et al. A role for tumor necrosis factor receptor 2 (TNFR-2) and receptor-interacting protein (RIP) in programmed necrosis and anti-viral responses. J Biol Chem 2003; 278:51613-51621.

35. Siegel RM, Chan FK, Chun HJ et al. The multifaceted role of Fas signaling in immune cell homeostasis and autoimmunity. Nat Immunol 2000; 1:469-474.

36. Krammer PH. CD95's deadly mission in the immune system. Nature 2000; 407:789-795.

37. Rieux-Laucat F, Le Deist F, Fischer A. Autoimmune lymphoproliferative syndromes: genetic defects of apoptosis pathways. Cell Death Differ 2003; 10:124-133.

38. Yeh WC, Itie A, Elia AJ et al. Requirement for Casper (c-FLIP) in regulation of death receptor-induced apoptosis and embryonic development. Immunity 2000; 12:633-642.

39. Yeh WC, Pompa JL, McCurrach ME et al. FADD: Essential for embryo development and signaling from some, but not all, inducers of apoptosis. Science 1998; 279:1954-1958.

40. Zhang J, Cado D, Chen A et al. Fas-mediated apoptosis and activation-induced T-cell proliferation are defective in mice lacking FADD/Mort1. Nature 1998; 392:296-300.

41. Varfolomeev EE, Schuchmann M, Luria V et al. Targeted disruption of the mouse Caspase 8 gene ablates cell death induction by the TNF receptors, Fas/Apo1, and DR3 and is lethal prenatally. Immunity 1998; 9:267-276.

42. Yeh WC, Hakem R, Woo M et al. Gene targeting in the analysis of mammalian apoptosis and TNF receptor superfamily signaling. Immunol Rev 1999; 169:283-302.

43. Lens SM, Kataoka T, Fortner KA et al. The caspase 8 inhibitor c-FLIP(L) modulates T-cell receptor-induced proliferation but not activation-induced cell death of lymphocytes. Mol Cell Biol. 2002; 22:5419-5433.

44. Tai TS, Fang LW, Lai MZ. c-FLICE inhibitory protein expression inhibits T-cell activation. Cell Death Differ 2004; 11:69-79.

45. Van Parijs L, Refaeli Y, Abbas AK et al. Autoimmunity as a consequence of retrovirus-mediated expression of C- FLIP in lymphocytes. Immunity 1999; 11:763-770.

46. Walsh CM, Wen BG, Chinnaiyan AM et al. A role for FADD in T cell activation and development. Immunity 1998; 8:439-449.

47. Newton K, Harris AW, Bath ML et al. A dominant interfering mutant of FADD/MORT1 enhances deletion of autoreactive thymocytes and inhibits proliferation of mature T lymphocytes. Embo J 1998; 17:706-718.

48. Newton K, Kurts C, Harris AW et al. Effects of a dominant interfering mutant of FADD on signal transduction in activated T cells. Curr Biol 2001; 11:273-276.

49. Salmena L, Lemmers B, Hakem A et al. Essential role for caspase 8 in T-cell homeostasis and T-cell-mediated immunity. Genes Dev 2003; 17:883-895.

50. Chun HJ, Zheng L, Ahmad M et al. Pleiotropic defects in lymphocyte activation caused by caspase-8 mutations lead to human immunodeficiency. Nature 2002; 419:395-399.
51. Zornig M, Hueber AO, Evan G. p53-dependent impairment of T-cell proliferation in FADD dominant- negative transgenic mice. Curr Biol 1998; 8:467-470.
52. Newton K, Harris AW, Strasser A. FADD/MORT1 regulates the preTCR checkpoint and can function as a tumour suppressor. Embo J 2000; 19:931-941.
53. Algeciras-Schimnich A, Griffith TS, Lynch DH et al. Cell cycle-dependent regulation of FLIP levels and susceptibility to Fas-mediated apoptosis. J Immunol 1999; 162:5205-5211.
54. Kirchhoff S, Muller WW, Krueger A et al. TCR-Mediated Up-Regulation of c-FLIP(short) Correlates with Resistance Toward CD95-Mediated Apoptosis by Blocking Death-Inducing Signaling Complex Activity. J Immunol 2000; 165:6293-6300.
55. Inaba M, Kurasawa K, Mamura M et al. Primed T cells are more resistant to Fas-mediated activation-induced cell death than naive T cells. J Immunol 1999; 163:1315-1320.
56. Kiener PA, Davis PM, Starling GC et al. Differential induction of apoptosis by Fas-Fas ligand interactions in human monocytes and macrophages. J Exp Med 1997; 185:1511-1516.
57. Perlman H, Pagliari LJ, Georganas C et al. FLICE-inhibitory protein expression during macrophage differentiation confers resistance to fas-mediated apoptosis. J Exp Med 1999; 190:1679-1688.
58. Willems F, Amraoui Z, Vanderheyde N et al. Expression of c-FLIP(L) and resistance to CD95-mediated apoptosis of monocyte-derived dendritic cells: Inhibition by bisindolylmaleimide. Blood 2000; 95:3478-3482.
59. Rescigno M, Piguet V, Valzasina B et al. Fas engagement induces the maturation of dendritic cells (DCs), the release of interleukin (IL)-1beta, and the production of interferon gamma in the absence of IL-12 during DC-T cell cognate interaction: a new role for Fas ligand in inflammatory responses. J Exp Med 2000; 192:1661-1668.
60. Micheau O. Cellular FLICE-inhibitory protein: An attractive therapeutic target? Expert Opin Ther Targets 2003; 7:559-573.
61. Semra YK, Seidi OA, Sharief MK. Overexpression of the apoptosis inhibitor FLIP in T cells correlates with disease activity in multiple sclerosis. J Neuroimmunol 2001; 113:268-274.
62. Sharief MK. Increased cellular expression of the caspase inhibitor FLIP in intrathecal lymphocytes from patients with multiple sclerosis [In Process Citation]. J Neuroimmunol 2000; 111:203-209.
63. Conlon P, Oksenberg JR, Zhang J et al. The immunobiology of multiple sclerosis: An autoimmune disease of the central nervous system. Neurobiol Dis 1999; 6:149-166.
64. Perlman H, Pagliari LJ, Liu H et al. Rheumatoid arthritis synovial macrophages express the Fas-associated death domain-like interleukin-1beta-converting enzyme-inhibitory protein and are refractory to Fas-mediated apoptosis. Arthritis Rheum 2001; 44:21-30.
65. Stassi G, Di Liberto D, Todaro M et al. Control of target cell survival in thyroid autoimmunity by T helper cytokines via regulation of apoptotic proteins. Nature Immunology 2000; 1:483-488.
66. Imanishi T, Murry CE, Reinecke H et al. Cellular FLIP is expressed in cardiomyocytes and down-regulated in TUNEL-positive grafted cardiac tissues. Cardiovasc Res 2000; 48:101-110.
67. Steenbergen C, Afshari CA, Petranka JG et al. Alterations in apoptotic signaling in human idiopathic cardiomyopathic hearts in failure. Am J Physiol Heart Circ Physiol Jan 2003; 284:H268-276.
68. Imanishi T, McBride J, Ho Q et al. Expression of cellular FLICE-inhibitory protein in human coronary arteries and in a rat vascular injury model. Am J Pathol 2000; 156:125-137.
69. Sata M, Walsh K. Endothelial cell apoptosis induced by oxidized LDL is associated with the down-regulation of the cellular caspase inhibitor FLIP. J Biol Chem 1998; 273:33103-33106.
70. Djerbi M, Screpanti V, Catrina AI et al. The inhibitor of death receptor signaling, FLICE-inhibitory protein defines a new class of tumor progression factors. J Exp Med 1999; 190:1025-1032.
71. Medema JP, de Jong J, van Hall T et al. Immune escape of tumors in vivo by expression of cellular FLICE- inhibitory protein. J Exp Med 1999; 190:1033-1038.
72. Screpanti V, Wallin RP, Ljunggren HG et al. A central role for death receptor-mediated apoptosis in the rejection of tumors by NK cells. J Immunol 2001; 167:2068-2073.
73. Taylor MA, Chaudhary PM, Klem J et al. Inhibition of the death receptor pathway by cFLIP confers partial engraftment of MHC class I-deficient stem cells and reduces tumor clearance in perforin-deficient mice. J Immunol 2001; 167:4230-4237.

74. Ryu BK, Lee MG, Chi SG et al. Increased expression of cFLIP(L) in colonic adenocarcinoma. J Pathol 2001; 194:15-19.
75. Thomas RK, Kallenborn A, Wickenhauser C et al. Constitutive expression of c-FLIP in Hodgkin and Reed-Sternberg cells. Am J Pathol 2002; 160:1521-1528.
76. Hueber AO, Zornig M, Bernard AM et al. A dominant negative Fas-associated death domain protein mutant inhibits proliferation and leads to impaired calcium mobilization in both T-cells and fibroblasts. J Biol Chem 2000; 275:10453-10462.
77. Wang J, Zheng L, Lobito A et al. Inherited human Caspase 10 mutations underlie defective lymphocyte and dendritic cell apoptosis in autoimmune lymphoproliferative syndrome type II. Cell 1999; 98:47-58.
78. Perlman H, Liu II, Georganas C et al. Differential expression pattern of the antiapoptotic proteins, Bcl-2 and FLIP, in experimental arthritis. Arthritis Rheum 2001; 44:2899-2908.

CHAPTER 5

Fas-Induced Necrosis

Tom Vanden Berghe, Nele Festjens, Michael Kalai, Xavier Saelens and Peter Vandenabeele

Abstract

Fas/CD95 is an important regulator of cell death in development and homeostasis of the immune system. Apoptosis is the most frequently observed type of cell death induced by Fas. It is characterized by cell shrinkage and nuclear fragmentation, while organelles and the plasma membrane retain their integrity for a prolonged period. Intensive studies of apoptotic cell death led to the discovery of the involvement of caspases. The first reports on necrotic caspase-independent cell death induced by Fas appeared in the late nineties. Necrotic cell death is characterized by minor nuclear changes and swelling of the cell, resulting in plasma and organelle membrane rupture. The current review focuses on Fas-initiated signaling events that allow a switch between apoptotic and necrotic cell death and on the mitochondrial processes that regulate an apoptotic or necrotic outcome. Finally, we describe events that are crucial for the execution of the necrotic cell death process.

Abbreviations

ACAD, activated T cell autonomous death; AICD, activation-induced cell death; ANT, adenine nucleotide translocator; Apaf-1, apoptotic protease activating factor-1; APC, antigen presenting cell; ATP, adenosine triphosphate; CA, cornu ammonis; CAD, caspase-activated DNA nuclease; cPLA2, Ca^{2+}-dependent phospholipase A2; dATP, deoxy adenosine 5'-triphosphate; DD, death domain; DED, death effector domain; DISC, death inducing signaling complex; ER, endoplasmic reticulum; FADD, Fas associated death domain; FasL, Fas ligand; HMGB1, high mobility group 1; Hsp, heat shock protein; IKK, inhibitory κB kinase; IMM, inner mitochondrial membrane; iPLA2, Ca^{2+}-independent phospholipase A2; LMP, lysosomal membrane permeability, MMP, mitochondrial membrane potential; MnSOD, Mn-superoxide dismutase; MPT, mitochondrial permeability transition; MPTP, mitochondrial permeability transition pore; MSDH, O-methyl-serine dodecylamide hydrochloride; NAD, nicotinamide-adenine dinucleotide; OMM, outer mitochondrial membrane; PARP, poly (ADP-ribose) polymerase; PCD, programmed cell death; PLAD, preligand association domain; PT, permeability transition; RIP1, receptor interacting protein 1; ROS, reactive oxygen species; SM, sphingomyelin; tBid, truncated Bid; TNF, tumor necrosis factor; TNFR, tumor necrosis factor receptor; TRAF2, TNF receptor associated factor 2; TRAIL-R1, TNF-related apoptosis-inducing ligand receptor; VDAC, voltage dependent anion channel.

Fas Signaling, edited by Harald Wajant. ©2006 Landes Bioscience and Springer Science+Business Media.

Three Types of Programmed Cell Death

The term programmed cell death (PCD) was initially introduced by R.A. Lockshin and C.M. Williams to designate cell death that occurred in a predictable place and time during embryonic development.[81] Schweichel and Merker distinguished three types of cell death based on morphology and the impact they have on neighboring cells in the developing embryo of mouse and rat and whether there was exposure to toxic compounds.[107] Literally quoted from the abstract of that paper: "Three distinct types of cell death were distinguished in control tissues. (1) Condensation and fragmentation of single cells undergoing phagocytosis, with lysosomal disintegration of the fragments in neighboring cells. (2) Primary formation of lysosomes in dying cells, with activation and subsequent destruction and phagocytosis of the fragments by neighboring cells. This type of cell death in most instances was found during destruction of organs or large cell units. (3) Disintegration of cells into fragments, which were optically no longer detectable, without involvement of the lysosomal system, e.g., in embryonic and epiphyseal cartilage before ossification." There are two interesting implications of this original work: three types of cell death seem to occur during normal ontogeny, and different toxic treatments of embryos were dominantly associated with one particular mode of cell death. Substances that disturbed replication, transcription or translation mainly caused type I apoptotic cell death, vitamin A caused type II autophagic cell death, and Mg-deficiency caused massive areas of type III necrotic cell death.[107] These observations led later on to the suggestion that depending on the stimulus and the cellular context one distinct cell death program will ensue, most probably because every cell death program is the result of signals that propagate that program while suppressing the others.[18] In brief, in type I cell death the distinct morphological changes include cell shrinkage and extensive chromatin condensation, in type II cell death there is formation of autophagic vacuoles inside the dying cell, and type III cell death is distinguished by rapid loss of plasma membrane integrity and spillage of the intracellular contents.[17]

Fas Is an Important Regulator of Cell Death

Fas/CD95 was originally identified as a cell surface antigen during attempts to characterize cell surface molecules involved in growth control of malignant lymphocytes.[120,140] Surprisingly, a monoclonal antibody specific for Fas proved highly cytotoxic to lymphocytes. Subsequent molecular cloning revealed that Fas is a type I transmembrane protein with three extracellular cysteine-rich domains, the hallmark of the tumor necrosis factor (TNF) receptor superfamily.[52,110] The cytotoxic activity of Fas depends on a conserved part of the cytoplasmic domain called death domain (DD) and therefore Fas is typically referred to as a death receptor.[51] It was observed that Fas-induced death by apoptosis, characterized by cell shrinkage and nuclear fragmentation, while organelles and the plasma membrane retained their integrity for a prolonged period.[52] Intensive studies of Fas-induced apoptosis led to the discovery of the involvement of cysteine aspartate-specific proteases, called caspases, in the apoptotic cell death process.[10,29,40,83] It has become clear that caspases comprise an evolutionarily conserved family of cysteine proteases involved in apoptotic signal transduction and processing of inflammatory cytokines (eleven human and ten mouse caspases).[69] However, in the late nineties, the first reports illustrating the capacity of Fas to initiate caspase-independent necrotic cell death appeared.[130]

Fas-Induced Apoptosis

Fas is expressed on the cell surface as preassociated homotrimers due to presence of a pre-ligand binding assembly domain (PLAD) at the N-terminus of the extracellular part of the receptor.[98,109] Upon ligand binding with Fas ligand (FasL) or agonistic antibodies, microaggregates of Fas form a complex known as the death-inducing signaling complex (DISC) with concomitant internalization into an endosomal pathway.[4,62] It has been suggested that ceramide and lipid rafts are involved in clustering and internalization of Fas, although this remains debatable and may be cell type dependent.[4,20,34,49,73] The physiological implication of Fas internalization is still unknown. This internalization may render the cells refractory to the cytotoxic effects of Fas-ligand expressed by neighboring cells or it may play a role in the escape of tumor cells from Fas-mediated apoptosis.[99] Additionally, internalization could be involved in signaling by promoting formation of the DISC.

In the DISC, the adaptor molecule Fas-associated death domain containing protein (FADD) is bound to Fas through homotypic interaction between its C-terminal DD and the C-terminal DD of Fas.[16,45,62] However, residues of the N-terminal death effector domain (DED) of FADD also seem to contribute to the interaction between FADD and Fas.[119] Through its N-terminal DED, FADD recruits other proteins, such as procaspase-8, also called FLICE(FADD-homologous ICE/CED-3-like protease) or MACH (MORT1/FADD-interacting protease), caspase-10 and c-FLIP.[10,93] Other proteins such as Daxx, FAP-1, FLASH, FAF1 and DAP3 were also reported to be recruited to the DISC and are reviewed in Peter and Krammer, 2003.[99] The recruitment of procaspase-8 by FADD provides a platform on which procaspase-8 undergoes a conformational change resulting in its proteolytic activation and the release of the active enzyme from the DISC complex as a p20/p10 heterotetramer.[25] Depending on the level of caspase-8 activation in the DISC and the susceptibility to cell death inhibition by Bcl-2 overexpression upon death receptor activation, two types of cells can be distinguished.[4,103,104] In type I cells, DISC formation and concomitant caspase-8 activation result in direct activation of downstream executioner caspases and apoptosis which cannot be blocked by Bcl-2 overexpression.[87,114] In type II cells, however, procaspase-8 and -3 activation is delayed due to greatly reduced DISC formation, resulting in dependence on the release of mitochondrial proteins to propagate and amplify the apoptotic signal. A molecular link connecting DISC activation and mitochondria is provided by caspase-8 mediated cleavage of Bid, generating tBid.[77] tBid translocates to the mitochondria and induces the release of mitochondrial factors, one of which is cytochrome c.[123] The latter triggers the assembly of the apoptosome, a high molecular weight caspase activating complex composed of apoptosis protease-activating factor 1 (Apaf-1), dATP and cytochrome c.[2,78] Bcl-2 overexpression inhibits receptor-induced apoptosis of type II cells as seen, for example in hepatic apoptosis induced by agonistic anti-Fas antibodies.[68] However, the physiological relevance of type II cells in the case of Fas-mediated apoptosis has been questioned, the argument being that results are often based on the use of agonistic Fas antibodies that do not necessarily mimic proper Fas receptor activation by Fas ligand.[48] It is still unclear what molecular mechanisms determine the efficiency of DISC formation and the distinction between type I and type II responses.

Decision Points in Fas-Induced Apoptosis or Necrosis

In this section we focus on the different levels of regulation that contribute to the final cell death outcome being apoptosis or necrosis. We discuss data that are in support of FADD as a platform for the initiation of either apoptotic or necrotic cell death. Next, we discuss the importance of expression of pro- and anti-apoptotic Bcl-2-members, the availability of ATP, the generation of reactive oxygen species (ROS), and the concentration of cytosolic calcium for the initiation of necrotic cell death at the level of mitochondria.

Necrosis or Apoptosis Decided at the Death Receptor Complex

We have demonstrated that addition of caspase-inhibitors blocks Fas-induced apoptosis in L929 murine fibrosarcoma cells without preventing cell death. Instead, the cells die by a necrotic process characterized by an intact nucleus and cytoplasmic swelling.[130] Necrotic cell death does not lead to tBid generation, the release of mitochondrial proteins or the generation of hypoploid DNA.[24] Similarly, Fas triggering in Jurkat cells in the presence of zVAD-fmk or

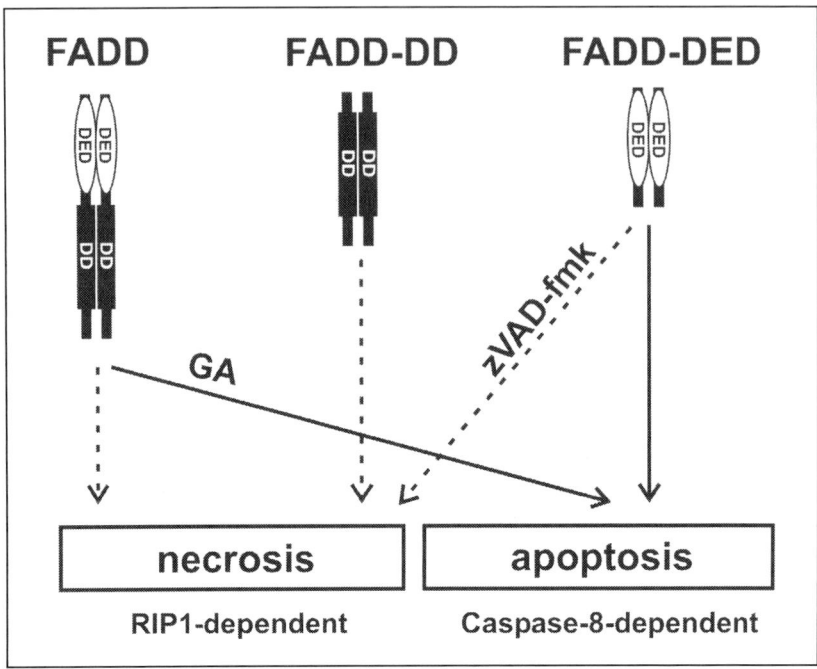

Figure 1. Schematic overview of the differential FADD-mediated signaling to cell death. Enforced dimerization of FKBP-FADD and FKBP-FADD-DD leads to necrosis, while FKBP-FADD-DED leads to apoptosis. The induction of necrosis by FKBP-FADD, FKBP-FADD-DD and FKBP-FADD-DED (in the presence of zVAD-fmk) may be explained by the ability of both to recruit RIP1. In the absence of caspase-inhibitors FKBP-FADD-DED induces apoptosis due to preferential recruitment of procaspase-8, competing out RIP1. FKBP-FADD-induced necrosis can be reverted to apoptosis by pretreating the cells with the Hsp90-inhibitor geldanamycin, a treatment that leads to the degradation of Hsp90-binding client proteins such as RIP1. FKBP-FADD-DD cannot recruit procaspase-8, explaining the inability of FKBP-FADD-DD to shift the response from necrosis to apoptosis in the presence of geldanamycin.

in Jurkat cells deficient in caspase-8 results in necrotic cell death.[56] In addition, enforced oligomerization of (FKBP)$_2$-FADD in these caspase-8-deficient Jurkat cells or in wild-type Jurkat cells treated with zVAD-fmk, results in necrotic cell death.[56] The same stimuli in wild-type Jurkat cells in the absence of caspase-inhibitors lead to apoptosis. These results are in agreement with the observation that overexpression of a FADD dominant-negative mutant, lacking the DED and thus unable to recruit and activate caspase-8, induces TNF-mediated necrosis in U937, NIH3T3[58] and L929 cells.[12]

Taken together these results suggest that the absence or inhibition of caspase-8 promotes necrotic cell death and that the death domain of FADD is able to propagate caspase-independent necrotic cell death. So FADD is able to initiate two cell death pathways: one through its DED requiring caspase-8 and leading to apoptotic cell death and the other requiring the DD and resulting in caspase-independent necrotic cell death (Fig. 1).[12,126] Which effector molecules propagate necrotic cell death downstream of the death domain of FADD? Studies in Jurkat cells revealed that receptor interacting protein 1 (RIP1), a DD-containing Ser/Thr kinase involved in TNF-mediated activation of NF-κB and MAPKs, is a mediator of Fas-mediated necrosis.[46] If caspase-activity is blocked, RIP1-deficient Jurkat cells are resistant to both types of cell death. Reintroducing RIP1 restores a necrotic cell death response in conditions under

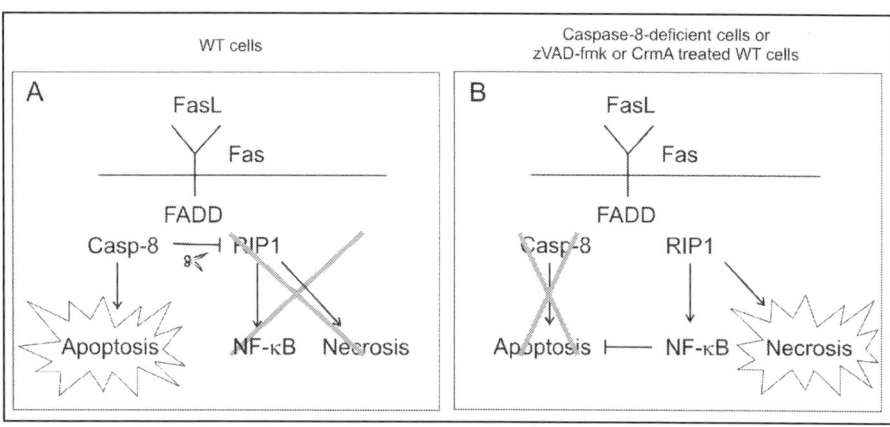

Figure 2. Schematic overview of the differential Fas-signaling to cell death. In Fas-induced signaling, FADD seems to function as a platform to initiate both apoptosis and necrosis. A) In WT cells the recruitment of procaspase-8 by FADD provides a platform on which procaspase-8 undergoes a conformational change resulting in its catalytic activation and apoptotic cell death. Interestingly, RIP1 as a key mediator in the necrotic cell death pathway and anti-apoptotic pathways through the activation of NF-κB, is a direct substrate of caspase-8. B) Fas stimulation in cells, in which caspase-8 is inhibited by zVAD-fmk or CrmA, or in caspase-8-deficient cells leads to necrotic cell death. In this condition, RIP1 is not cleaved and its recruitment to the receptorcomplex leads to anti-apoptotic signaling through the activation of NF-κB, and necrotic cell death through its kinase-activity. This suggests that apoptotic and necrotic cell death pathways are interrelated from the initiation at the death receptor level. Successful activation of caspase-8 will result in suppression of both the necrotic and anti-apoptotic pathways initiated by RIP1. The absence of caspase-8, allows RIP1 recruitment favoring necrotic and anti-apoptotic pathways.

which caspase activity is blocked. Moreover, the kinase activity of RIP1 is required for rescuing the necrotic cell death response.[46] Indirect evidence for an essential role of RIP1 in transducing DD-receptor induced necrosis came from studies on heat shock protein 90 (Hsp90). Hsp90 is a cytosolic chaperone for many kinases, including RIP1.[76] Inhibition of Hsp90 by geldanamycin or radicicol, a structurally unrelated inhibitor, leads to a 10-fold downregulation of RIP1 levels and inhibits Fas- and TNFR1-induced necrosis.[46,125] Moreover, pretreatment of L929 cells with geldanamycin results in an apoptotic response to TNF. It should be noted that other kinases, such as IKK-α and -β are also down regulated by Hsp90 inhibition providing a cellular content that propagates apoptotic cell death instead of necrotic cell death.[125]

Can FADD recruit RIP1? In a yeast two hybrid screening we found that FADD was able to bind RIP1 (Maria van Gurp, unpublished results).[128] Using transient overexpression and immunoprecipitation in HEK293 cells, we confirmed that FADD and FADD-DD were able to recruit RIP1. To our surprise FADD-DED was also able to recruit overexpressed RIP1. However, cotransfection of procaspase-8 C/A reduced the levels of FADD-DED-mediated recruitment of RIP1 to background levels.[126] The interaction between FADD-DED and RIP1 may explain why induction and dimerization of FADD-DED in the presence of caspase-inhibitors can still initiate necrotic cell death (Fig. 1). It is interesting that RIP1, a key mediator in the necrotic cell death pathway and anti-apoptotic pathways that functions by activation of NFκB, is also a direct substrate of caspase-8.[61,80,84] In this model, successful activation of caspase-8 may result in suppression of the necrotic and anti-apoptotic pathways initiated by RIP1 (Fig. 2). This suggests that apoptotic and necrotic cell death pathways not only compete at the site of initiation of the signal but also interfere with each other.

Another possible switch between apoptotic and necrotic cell death starting from Fas is provided by the extent of ceramide generation. Ceramide is derived from sphingomyelin (SM), initially located in the outer leaflet of the plasma membrane. SM gains access to a cytosolic sphingomyelinase by flipping to the inner leaflet in a process of lipid scrambling.[118] Scrambling-deficient Raji B cells do not show ceramide formation and favor apoptosis, while cells that exhibit Fas-mediated ceramide generation such as Jurkat cells and A20 B cells display both apoptotic and necrotic cell death features.[42] Moreover, addition of cell-permeable C6- or C2-ceramide was shown to induce cell death without caspase-3 activation, DNA fragmentation, cell shrinkage and chromatin condensation. Instead, cells increased in size and became filled with vacuolar structures, features that resemble FasL-induced necrosis.[42] These observations illustrate another level of regulation in which there is an association between Fas-mediated necrosis, lipid scrambling and late ceramide production.

Although the molecular mechanisms during TNFR1- and Fas-induced apoptosis have mostly been presented as essentially similar, recent publications have pinpointed some clear distinctions.[39,88] According to these publications TNFR1-induced apoptosis involves two sequential signaling complexes. The initial plasma membrane bound complex (complex I) consists of TNFR1, the adaptor TRADD, the kinase RIP1 and TRAF2. This complex rapidly activates the IKK complex. Due to unknown modifications complex I dissociates from the receptor and a cytosolic complex (complex II) is formed. Complex II consists of TRADD, RIP1, TRAF2, FADD and procaspase-8.[88] In contrast, Fas, TNF-related apoptosis-inducing ligand receptor-1 (TRAIL-R1) and TRAIL-R2 bind FADD and caspase-8 directly.[39]

Although the kinase activity of RIP1 is obligatory for TNFR1-induced necrosis, FADD seems to be dispensable. In TNFR1-induced necrosis, FADD is dispensable for the recruitment of RIP1, since RIP1 can interact directly with TNFR1.[39] and/or TRADD.[47] Remarkably, in FADD-deficient Jurkat cells RIP1 is also recruited to Fas, but no toxicity is observed after Fas triggering,[46] demonstrating that FADD is indispensable for both Fas-mediated apoptosis and necrosis. In contrast, FADD-deficiency renders Jurkat cells even more sensitive to TNFR1-induced necrotic cell death.[39,46] This suggests that a FADD-interacting protein has an anti-necrotic effect. Indeed, overexpression of CrmA, a potent caspase-8 inhibitor, sensitizes cells to TNF-induced necrosis.[129] Therefore, it is conceivable that the presence of caspase-8 negatively influences the initiation of necrotic cell death.

The requirement for the kinase activity of RIP1 in necrotic cell death induced by Fas ligand raises the question: what are the targets of RIP1? Since production of reactive oxygen species by oxidative phosphorylation has been reported as an important process in the execution of necrotic cell death,[33,106] it is possible that RIP1 directly or indirectly targets mitochondria. On the other hand, in view of the possible role of cathepsins in necrotic cell death,[117] it is also possible that RIP1 directly or indirectly targets the lysosomes. A possible link is provided by the activation of calpains[115,116] and phospholipase A2[63,102] during necrotic cell death, since it has been suggested that calpains[139] and PLA2[143] both destabilize lysosomal integrity.

Necrosis or Apoptosis Decided at the Mitochondrial Level

In several experimental setups mitochondria were shown to act as a rheostat, controlling the ratio between apoptotic and necrotic outcome. The onset of necrosis is associated with hyperproduction of ROS, ATP-depletion and increased cytosolic calcium. In addition, the expression level of several Bcl-2 family members may also determine the decision between apoptotic and necrotic cell death.

Role of ATP-Levels

Mitochondria are known to be endowed with a latent mechanism called the mitochondrial permeability transition (MPT), involving the opening of a nonspecific pore, known as the mitochondrial permeability transition pore (MPTP), formed at the contact sites between the inner mitochondrial (IMM) and the outer mitochondrial membranes (OMM).[21] Under normal physiological conditions, the opening of the pore is well regulated and functions predominantly in the exchange of ADP for ATP, mediated by the adenine nucleotide translocator (ANT), which is, next to the voltage dependent anion channel (VDAC), a core component of the pore. VDAC in turn makes the OMM permeable to most small molecules (<5kD), allowing free exchange of respiratory chain substrates. However, when MPT is induced, the MPTP is constitutively opened with consequent IMM permeabilization. This event results in loss of matrix components, and impairment of mitochondrial functionality due to loss of the transmembrane potential, which is required for proper oxidative phosphorylation.

The eventual type of cell death depends on several factors including the availability of ATP.[21,59,127] If the onset of the MPT is widespread and impairs most of the mitochondria in the cell, mitochondrial oxidative phosphorylation and ATP regeneration is affected.[78] If on top of that glycolytic sources of ATP are unavailable, the cell becomes profoundly depleted of ATP and finally there prevails a caspase-independent death resembling necrosis, as shown for ischemic rat hepatocytes and calcium ionophore-induced cell death.[26,59,60] Reperfusion in the presence of the glycolytic substrate fructose prevents necrotic killing of hepatocytes after simulated ischemia/reperfusion although the onset of MPT and caspase- and ATP-dependent apoptosis occurs.[60] The same holds true for calcium ionophore-induced cell death, which is apoptotic when ATP sources are available.[26] ATP-dependent steps of Fas-mediated apoptotic signal transduction have also been studied.[27,30] In type I cells, ATP-dependent steps of Fas-induced apoptosis are located downstream of caspase-3 activation, and probably include an active machinery of nuclear changes. In type II cells however, activation of the apoptosome is the most upstream ATP-dependent step. Apoptosis of Jurkat cells elicited by anti-Fas antibodies in glucose-free medium is converted to necrosis in the presence of NO donors, which induce failure of mitochondrial energy production.[75] ATP depletion also enhances Fas-mediated hepatotoxicity, which provides an explanation for the occurrence of relatively widespread necrosis in pathological liver failure where ATP and glutathione are frequently depleted.[72] Moreover, *Apaf-1* knockout mice exhibit programmed necrotic cell death of the interdigital cells instead of apoptotic cell death, demonstrating that the inability to form the ATP-dependent apoptosome favors necrotic cell death in vivo.[15]

Beside massive ATP depletion through dysfunction of the mitochondria, ATP levels are also greatly reduced by continuous activation of poly-ADP ribose polymerase (PARP), an enzyme that participates in the detection and repair of single strand DNA breaks.[134] DNA damaging stimuli such as ionizing and UV radiation, topoisomerase I inhibitors and alkylating agents cause PARP activation, stimulating DNA repair and permitting cell survival. However, overactivation of PARP was suggested to lead to the depletion of its substrate NAD^+, thus slowing the rates of glycolysis, electron transport and consequent ATP formation.[37,131] Blocking PARP activity by caspase-mediated cleavage was therefore suggested to be vital for the preservation of the cellular energy needed for certain ATP-sensitive steps during the apoptotic process, and guarantees proper functioning of the apoptotic machinery. Experimental evidence supporting this hypothesis was provided by Herceg and Wang, who demonstrated that the expression of a caspase-uncleavable form of PARP in TNF-treated PARP-1 knockout fibroblasts leads to NAD^+ and ATP depletion and necrosis.[41,82] However, it remains likely that the main role of the cleavage and inactivation of PARP is to prevent DNA repair.

Role of Pro- and Anti-Apoptotic Members of the Bcl-2-Family

A necrotic outcome subsequent to MPT does not result only from the depletion of cellular ATP levels. The eventual type of cell death may also depend on the expression level of pro- versus anti-apoptotic members of the Bcl-2 family, the crucial integrators of survival and death signals in higher eukaryotes.[19,31] The importance of these sentinels of cellular integrity was shown for hyperoxia-induced necrotic cell death in A549 human alveolar epithelial cells.[133] The induction of cell death under these conditions involves the formation of the DISC complex, and was illustrated through the interaction of caspase-8 with Fas in immunoprecipitation experiments. Despite the initiation of an apoptotic program and sustained levels of ATP, cells eventually died by necrosis. In A549 cells subjected to hyperoxia, an increased expression, activation and mitochondrial translocation of Bid and Bax were observed, together with an increased expression of Bcl-x_L without a change in the level of Bcl-2. Despite this increased expression level, Bcl-x_L could not protect against hyperoxic lung injury and final necrotic cell death in vitro or in vivo. These results are in line with an earlier report describing how Bcl-x_L could not prevent Bax-induced mitochondrial damage and apoptosis when expressed in low levels, but caused a Bax-induced necrotic outcome when highly expressed.[108] Similarly, lymphocytes expressing low levels of Bcl-2 die by apoptosis elicited by oxidized low-density lipoproteins or the calcium ionophore A23187. However, the same stimuli induced necrotic cell death of lymphocytes expressing high Bcl-2-levels.[86] These results demonstrate that the final cell fate is determined by a discrete interplay between Bcl-2 survival and death factors. Bcl-x_L and Bcl-2 are anti-apoptotic when normally expressed. However, if their expression level drops below a certain limit, they are unable to antagonize pro-apoptotic family members. Conversely, Bcl-x_L and Bcl-2 may become pro-necrotic if their expression exceeds normal levels.

The decision to die by apoptosis or necrosis is determined not only by the expression level of the anti-apoptotic Bcl-2 family members. BNIP3 is a death promoting "BH3-only" protein that is loosely associated with the OMM in normal tissue. Following transient overexpression, BNIP3 integrates into the OMM in a BH3 domain-independent manner with a consequent loss of mitochondrial membrane potential, an increase in ROS production and induction of a caspase-independent, necrotic-like cell death.[124] Necrotic-like cell death of chronic lymphoma B cells following the activation of the thrombospondin receptor CD47, may also be mediated by BNIP3.[70,85] BNIP3 is bound to the unoccupied receptor and translocates to the OMM upon ligand or antibody binding. The contribution of BNIP3 was confirmed by antisense studies showing that downregulation of BNIP3 partially attenuates CD47-mediated necrotic cell death. BNIP3 expression is increased in many cell types during hypoxia, consistent with the presence of a hypoxia responsive element in the promoter of the BNIP3 gene.[36] Moreover, blockage of hypoxia-induced BNIP3 expression using antisense oligonucleotides against BNIP3 or inhibition of BNIP3 function through expression of a mutant dominant-negative form of the protein protects human embryonic kidney 293 cells from hypoxia-induced cell death.[65] In situ hybridization studies on different human tumors showed that BNIP3 is up-regulated in peri-necrotic regions of most of these tumors.[111] Hypoxic regions within solid tumors are often resistant to chemotherapy and radiation possibly due to the circumvention of BNIP3-mediated hypoxia-induced cell death of human epithelial cells by growth factor signaling.[65] Recent studies show that hypoxia alone is not a major stimulus for cell death but must be combined with acidosis.[67] Hypoxia-acidosis-associated cell death also seems to be mediated by BNIP3. Chronic hypoxia induces the expression of BNIP3 mRNA in cardiac myocytes and acidosis stabilizes BNIP3 protein and increases the association with mitochondria. In line with these observations, hypoxia-acidosis-induced cell death can be blocked by pretreatment with antisense BNIP3 oligonucleotides.[67]

Role of Reactive Oxygen Species

Reactive oxygen species (ROS) are produced during normal physiological cellular events and are involved in various biologic processes, including activation of gene expression, regulation of proliferative events, and cellular response to cytokines.[3,90,122] Although ROS production is important for the normal functioning of the cell, overproduction of ROS can lead to apoptotic or necrotic cell death.[66] The mitochondrion is the major source of ROS within the cell. Superoxide radicals (O_2^-) are generated in the intermembrane space by electron leakage to molecular oxygen during the transfer of electron from reduced ubiquinon (UQH2) in complex III to cytochrome c,[38,64] and in the matrix from an undefined site in complex I.[7,8] Under normal conditions, these radicals are removed by Mn-superoxide dismutase (MnSOD), which generates hydrogen peroxide (H_2O_2), which in turn can be reduced to H_2O by catalase and GSH peroxidase. However, in some physiological and pathophysiological conditions, the balance between free radicals and scavengers is disturbed and overproduction of the former can cause cellular damage.[66] ROS can exert deleterious actions on proteins like MPT pore constituents, promoting opening of the pore, and can inactivate caspases.[89] Moreover, free radicals can trigger single strand breakage of DNA, in turn activating the nuclear enzyme PARP, with concomitant depletion of NAD^+ and ATP.[131] Oxygen radicals can also interact with plasma- and organelle membranes leading to lipid peroxidation and consequent disturbance of membrane integrity.[113] In mitochondria, lipid oxidation may increase the destructive effects of MPT in cells, promoting necrotic cell death instead of apoptosis. In addition, ROS-mediated disturbance of lysosomal membrane[100] and plasma membrane integrity may also contribute to the necrotic process (Discussed below).

During anti-Fas-induced cell death of a neuroglioma cell line, both the caspase-dependent and ROS-dependent pathways are activated.[53] It was shown that activation of caspases and increased production of H_2O_2 might only lead to apoptosis, whereas increased levels of superoxide can cause both apoptosis and necrosis, suggesting a dual mode of anti-Fas-induced cell death. In line with this, overproduction of Cu/ZnSOD could block the apoptotic and necrotic pathways whereas catalase overproduction inhibited apoptosis and potentiated the necrotic pathway.[53] This result is supported by the reported potentiation of superoxide-mediated pulmonary damage in rats by excess catalase,[71] which suggests that the reaction of high concentrations of catalase with superoxide radicals might lead to the formation of toxic superoxide-catalase products. This idea is further supported by experiments showing that the same stimuli can cause either apoptosis or necrosis depending on their concentration and cellular redox status.[74] In L929sA cells dsRNA induces necrosis that involves high reactive oxygen species production. The antioxidant butylated hydroxyanisole protects cells from necrosis, but shifts the response to apoptosis.[54] These results clearly demonstrate the importance of a strict equilibrium between ROS and their scavengers in the cellular decision to die by necrosis or apoptosis.

Role of Calcium-Homeostasis

Different reports show the importance of mitochondrial calcium homeostasis in the regulation of both apoptotic and necrotic cell death.[97,101,144] Mitochondrial calcium uptake is mediated primarily via the uniporter[35] whereas calcium efflux from the mitochondria is driven by a Na^+/Ca^{2+} antiporter and the MPT.[50] Low extracellular $[Ca^{2+}]$ ($\leq 2mM$) triggers reduction in mitochondrial $[Ca^{2+}]$ and mitochondrial membrane potential, resulting in apoptosis. Necrotic death however occurs in the presence of excess extracellular $[Ca^{2+}]$ ($\geq 5mM$), which in turn increases cytosolic $[Ca^{2+}]$ with concomitant mitochondrial $[Ca^{2+}]$ overload inducing a final decrease in mitochondrial membrane potential.[144] A cytosolic $[Ca^{2+}]$ increase during necrosis can also be the result of calcium release from the endoplasmic reticulum (ER) or mitochondrial stores. Besides inducing calcium release from the mitochondria, increased cytosolic $[Ca^{2+}]$ can also activate phospholipases or the cysteine protease calpain, which may also mediate the necrotic process (discussed below).

Enzymatic Systems Involved in the Execution of Necrotic Events

The identification of executioner pathways in the necrotic cell death process is often hampered by the occurrence of many secondary events during necrotic disintegration of the cell. In this section, we describe necrotic events that were demonstrated to be crucial for the execution of the cell death process in several necrotic cell death systems. These events include major contributions of (1) caspase-independent proteolytic cascades that involve calpains and lysosomal cathepsins, (2) phospholipases and (3) ROS producing systems. So far, none of these was examined in Fas-initiated necrosis.

Calpain Mediated Release of Cathepsins

An increase in cytosolic $[Ca^{2+}]$ results either from calcium release from the ER or mitochondrial stores or from an influx of extracellular calcium. The release of calcium from ER calcium stores can determine necrosis, as shown by the requirement for calreticulin and regulators of calcium release in necrotic neuronal death in *C. elegans*.[135] The increase in cytosolic calcium may trigger a calpain-mediated cell death. Calpains are cysteine proteases that are localized in the cytosol as inactive precursors, which are activated upon cytosolic calcium increase.[132] Several reports illustrate the importance of calpains in necrotic cell death e.g, degenerative (necrotic-like) cell death in neurons of *C. elegans* requires the activity of the calcium-regulated CLP-1 and TRA-3 calpain proteases.[115,116] Necrosis of dystrophin-deficient muscle results from increases in $[Ca^{2+}]$ and causes the activation of calpains.[112] Research on cornu ammonis (CA) 1 neuronal death in monkeys provided direct evidence for calpain activation prior to ischemic neuronal death.[138] Immunoelectron microscopy revealed that activated μ-calpain is localized at the vacuolated or disrupted membrane of lysosomes.[136,139] Moreover, in the ischemia-vulnerable CA1 neurons the cathepsin B immunoreactivity was localized not only to the lysosomes but also to the perikarya and/or the neuropil of the CA1 sector, indicating lysosomal leakage of cathepsin B.[137] The lysosomal membrane permeability (LMP) does not seem to be a secondary response as inhibitors of cathepsins B and L significantly reduced neuronal death.[137,139] Accordingly, these results led to the definition of a "calpain-cathepsin hypothesis", suggesting a calpain mediated lysosomal disruption with the consecutive release of cathepsin B and L. Several other groups reported a role for cathepsin B in caspase-independent cell death by TNF family ligands. TWEAK can induce both caspase-dependent apoptosis and cathepsin B-dependent necrosis, in a cell type-specific manner.[94] TNF-induced cell death of hepatocytes and tumor cells apparently requires lysosomal cysteine protease cathepsin B.[32] Whether calpain and/or cathepsin activity is essential for Fas-induced necrosis is still unclear, but it is likely in view of their roles in necrotic cell death by other TNF family ligands.

Activation of Phospholipases

The involvement of phospholipases, and mainly phospholipases A2 (PLA2), in programmed cell death has been reported several times. PLA2s cleave a sn-2 ester bond in glycerophospholipids, releasing free fatty acids and lysophospholipids. The role of PLA2s in apoptosis and necrosis depends upon the cell type, the stimulus of injury and the PLA2 isoform.[22] Cytosolic, calcium-dependent PLA (cPLA2) seems to be required for TNF-induced apoptosis in mouse L929 cells whereas it is dispensable in Fas-mediated apoptosis suggesting that the TNF receptor and Fas use different signaling pathways for apoptosis.[23,28,43] Active caspase-3 generated during Fas-mediated apoptosis of human leukemic U937 cells, leads to proteolytic inactivation of cPLA2,[5] which would contribute to maintenance of plasma membrane integrity, a characteristic feature of an apoptotically dying cell. Furthermore, as cPLA2 activation is an essential step in the generation of proinflammatory lipid mediators,[92] its inactivation may be part of the anti-inflammatory feature of apoptosis. However, the inducible calcium-independent PLA2 (iPLA2) is not a target for caspase-mediated inactivation and may, if induced, compensate for the inactivation of cPLA2.[5,6] Cleavage of cPLA2 during caspase-independent necrotic cell death has not been reported yet. Moreover, cPLA2 seems to play a major role in chemical- or oxidant-induced renal necrosis.[63,102] It is conceivable that during Fas-induced necrosis, cPLA2-induced hydrolysis of phospholipids may disturb membrane integrity by acting as a detergent and altering membrane fluidity.

Lysosomal Membrane Permeability by Free Radicals

It is likely that not only calpain proteolysis but also mitochondrial oxidative stress is capable of provoking LMP leading to necrotic cell death. Oxidative stress e.g., xanthine or naphthazarin, induce the release of lysosomal enzymes both in vitro[55] and in situ.[96] As a result of the continuous digestion of iron-containing metalloproteins such as cytochromes and ferritin, the lysosomes within normal cells contain a pool of labile, redox-active Fe^{2+}, which make these organelles particularly susceptible to oxidative damage.[141] Oxidant-mediated destabilization of lysosomal membranes by an iron-mediated Fenton reaction provokes the release of hydrolytic enzymes in the cytoplasm which leads to a cascade of events eventually resulting in cell death by apoptosis or necrosis depending on the severity of the insult.[14,141] Addition of low concentrations of O-methyl-serine dodecylamide hydrochloride (MSDH), a lysosomotropic agent,[79] results in partial LMP, activation of caspases and apoptosis. At high concentrations, MSDH causes extensive LMP and necrosis. LMP can also activate MPT, and thus lead to cell death.[13]

Nuclear Events

Along with proteolysis, necrosis is often accompanied by degradation of DNA.[1] Degradation of DNA during necrosis usually occurs randomly, forming a smear of DNA on agarose gels, in contrast to apoptotic DNA which shows the characteristic ladder pattern on the gels.[44] Whether DNA is always fragmented remains a controversial issue. In our model system of TNF-induced necrosis we did not observe DNA fragmentation.[24] Others have shown that necrosis following renal ischemia/reperfusion[9] and brain ischemia[121] is associated with increased DNAse I and DNAse II activity, respectively. Recently it was demonstrated that extracellular DNAse I present in serum, which is not heat inactivated, contributes to DNA fragmentation following necrotic cell death.[95] Thus, extracellular DNAse I and lysosomal DNAse II are probably implicated in DNA fragmentation in necrosis, whereas the main apoptotic nuclease is CAD (caspase-activated DNase).[142] Of note is that CAD[-/-] thymocytes do not undergo DNA degradation, unless they are phagocytosed by macrophages,[57] demonstrating a contribution of lysosomal DNase II from the phagocytic cell to the degradation of DNA of the engulfed cell or the apoptotic body.

Recently high mobility group 1 (HMGB1) protein was discovered as a biochemical necrotic marker, as it is passively released by necrotic or damaged cells.[105] This release signals the demise of a cell to its neighbors. HMGB1 is both a nuclear factor and a secreted protein. In the nucleus it acts as an architectural chromatin-binding factor that bends DNA, stabilizes nucleosomes and facilitates transcription.[11] Outside the cell, it binds with high affinity to RAGE (the receptor for advanced glycation end products) and is a potent mediator of inflammation, cell migration and metastases.[91] Apoptotic cells do not release HMGB1 even when progressing to secondary necrosis and partial autolysis. The retention of HMGB1 by apoptotic cells may constitute a mechanism to avoid inflammation even if phagocytic clearance fails.[105]

Conclusions

Fas typically induces apoptotic cell death. However, when caspase-8 is absent or blocked a RIP1-dependent necrotic pathway prevails. This suggests that the composition of the death domain receptor complex determines the eventual cell death type. More downstream in the cell death signaling pathway, the mitochondria function as an additional rheostat, deciding the eventual cell fate as apoptosis or necrosis. A drastic drop in mitochondrial ATP production coinciding with a failure of glycolytic ATP production eventually leads to necrotic cell death. Increase in cytosolic calcium or an inappropriate balance between ROS production and radical scavengers are also cellular conditions that favor the necrotic cell death process. In addition, pro-apoptotic Bcl-2 members such as Bax and Bid, or other stimuli such as a calcium ionophore and LDL in the presence of low levels of anti-apoptotic Bcl-2, will induce caspase-dependent apoptosis, while high levels of Bcl-2 still allows a caspase-independent form of cell death that resembles necrosis, suggesting that the level of expression of anti-apoptotic Bcl-2 members determines the type of cell death outcome.

The link between Fas-mediated initiation of necrotic cell death in which both FADD and RIP1 seem to be crucial and the downstream necrotic executioner mechanisms remains unclear. Several reports argue for an implication of caspase-independent proteolytic cascades that involve calpains and lysosomal cathepsins, phospholipases, mitochondrial ROS production and effects of oxidative phosphorylation, and ROS-mediated destabilization of the lysosomal membrane. A major challenge now is to link these effector mechanisms with the initiation events. This will allow us to define the necrotic cell death pathway as an alternative form of PCD. Besides its role in pathologies, the physiological contribution of necrotic cell death in development and homeostasis remains obscure.

Acknowledgements

Research in the Molecular Signalling and Cell Death Unit was supported in part by the IUAP-V, the Fonds voor Wetenschappelijk Onderzoek—Vlaanderen (grant 3G.0006.01 and grant 3G.021199), an EC-RTD grant QLG1-CT-1999-00739, a RUG-cofinanciering EU project (011C0300), and a RUG-GOA project (12050502). XS, MK, and PV are paid by the GOA, EC-RTD and University Gent, respectively.

References

1. Abrahamse SL, van Runnard Heimel P, Hartman RJ et al. Induction of necrosis and DNA fragmentation during hypothermic preservation of hepatocytes in UW, HTK, and Celsior solutions. Cell Transplant 2003; 12:59-68.
2. Adams JM, Cory S. Life-or-death decisions by the Bcl-2 protein family. Trends Biochem Sci 2001; 26:61-6.
3. Adler V, Yin Z, Tew KD et al. Role of redox potential and reactive oxygen species in stress signaling. Oncogene 1999; 18:6104-11.
4. Algeciras-Schimnich A, Shen L, Barnhart BC et al. Molecular ordering of the initial signaling events of CD95. Mol Cell Biol 2002; 22:207-20.
5. Atsumi G, Tajima M, Hadano A et al. Fas-induced arachidonic acid release is mediated by Ca^{2+}-independent phospholipase A2 but not cytosolic phospholipase A2, which undergoes proteolytic inactivation. J Biol Chem 1998; 273:13870-7.
6. Balsinde J, Dennis EA. Distinct roles in signal transduction for each of the phospholipase A2 enzymes present in P388D1 macrophages. J Biol Chem 1996; 271:6758-65.
7. Barja G. Mitochondrial oxygen radical generation and leak: Sites of production in states 4 and 3, organ specificity, and relation to aging and longevity. J Bioenerg Biomembr 1999; 31:347-66.
8. Barja G, Herrero A. Localization at complex I and mechanism of the higher free radical production of brain nonsynaptic mitochondria in the short-lived rat than in the longevous pigeon. J Bioenerg Biomembr 1998; 30:235-43.

9. Basnakian AG, Ueda N, Kaushal GP et al. DNase I-like endonuclease in rat kidney cortex that is activated during ischemia/reperfusion injury. J Am Soc Nephrol 2002; 13:1000-7.

10. Boldin MP, Goncharov TM, Goltsev YV et al. Involvement of MACH, a novel MORT1/FADD-interacting protease, in Fas/APO-1- and TNF receptor-induced cell death. Cell 1996; 85:803-15.

11. Bonaldi T, Langst G, Strohner R et al. The DNA chaperone HMGB1 facilitates ACF/CHRAC-dependent nucleosome sliding. Embo J 2002; 21:6865-73.

12. Boone E, Vanden Berghe T, van Loo G et al. Structure/Function analysis of p55 tumor necrosis factor receptor and fas-associated death domain. Effect on necrosis in L929sA cells. J Biol Chem 2000; 275:37596-603.

13. Boya P, Andreau K, Poncet D et al. Lysosomal membrane permeabilization induces cell death in a mitochondrion-dependent fashion. J Exp Med 2003; 197:1323-34.

14. Brunk UT, Terman A. The mitochondrial-lysosomal axis theory of aging: Accumulation of damaged mitochondria as a result of imperfect autophagocytosis. Eur J Biochem 2002; 269:1996-2002.

15. Chautan M, Chazal G, Cecconi F et al. Interdigital cell death can occur through a necrotic and caspase-independent pathway. Curr Biol 1999; 9:967-70.

16. Chinnaiyan AM, Tepper CG, Seldin MF et al. FADD/MORT1 is a common mediator of CD95 (Fas/APO-1) and tumor necrosis factor receptor-induced apoptosis. J Biol Chem 1996; 271:4961-5.

17. Clarke PG. Developmental cell death: Morphological diversity and multiple mechanisms. Anat Embryol (Berl) 1990; 181:195-213.

18. Clarke PG. Apoptosis: From morphological types of cell death to interacting pathways. Trends Pharmacol Sci 2002; 23:308-9.

19. Cory S, Adams JM. The Bcl2 family: Regulators of the cellular life-or-death switch. Nat Rev Cancer 2002; 2:647-56.

20. Cremesti A, Paris F, Grassme H et al. Ceramide enables fas to cap and kill. J Biol Chem 2001; 276:23954-61.

21. Crompton M. The mitochondrial permeability transition pore and its role in cell death. Biochem J 1999; 341(Pt2):233-49.

22. Cummings BS, McHowat J, Schnellmann RG. Phospholipase A(2)s in cell injury and death. J Pharmacol Exp Ther 2000; 294:793-9.

23. De Valck D, Vercammen D, Fiers W et al. Differential activation of phospholipases during necrosis or apoptosis: A comparative study using tumor necrosis factor and anti-Fas antibodies. J Cell Biochem 1998; 71:392-9.

24. Denecker G, Vercammen D, Steemans M et al. Death receptor-induced apoptotic and necrotic cell death: Differential role of caspases and mitochondria. Cell Death Differ 2001; 8:829-40.

25. Donepudi M, Mac Sweeney A, Briand C et al. Insights into the regulatory mechanism for caspase-8 activation. Mol Cell 2003; 11:543-9.

26. Eguchi Y, Shimizu S, Tsujimoto Y. Intracellular ATP levels determine cell death fate by apoptosis or necrosis. Cancer Res 1997; 57:1835-40.

27. Eguchi Y, Srinivasan A, Tomaselli KJ et al. ATP-dependent steps in apoptotic signal transduction. Cancer Res 1999; 59:2174-81.

28. Enari M, Hug H, Hayakawa M et al. Different apoptotic pathways mediated by Fas and the tumor-necrosis-factor receptor. Cytosolic phospholipase A2 is not involved in Fas-mediated apoptosis. Eur J Biochem 1996; 236:533-8.

29. Enari M, Hug H, Nagata S. Involvement of an ICE-like protease in Fas-mediated apoptosis. Nature 1995; 375:78-81.

30. Ferrari D, Stepczynska A, Los M et al. Differential regulation and ATP requirement for caspase-8 and caspase-3 activation during CD95- and anticancer drug-induced apoptosis. J Exp Med 1998; 188:979-84.

31. Festjens N, Van Gurp M, van Loo G et al. Bcl-2 family members as sentinels of cellular integrity and role of mitochondrial intermembrane space proteins in apoptotic cell death. Acta Haematol 2004; 111:7-27.

32. Foghsgaard L, Wissing D, Mauch D et al. Cathepsin B acts as a dominant execution protease in tumor cell apoptosis induced by tumor necrosis factor. J Cell Biol 2001; 153:999-1010.

33. Goossens V, Grooten J, De Vos K et al. Direct evidence for tumor necrosis factor-induced mitochondrial reactive oxygen intermediates and their involvement in cytotoxicity. Proc Natl Acad Sci USA 1995; 92:8115-9.

34. Grassme H, Jekle A, Riehle A et al. CD95 signaling via ceramide-rich membrane rafts. J Biol Chem 2001; 276:20589-96.

35. Gunther T, Vormann J. Intracellular Ca(2+)-Mg2+ interactions. Ren Physiol Biochem 1994; 17:279-86.

36. Guo K, Searfoss G, Krolikowski D et al. Hypoxia induces the expression of the pro-apoptotic gene BNIP3. Cell Death Differ 2001; 8:367-76.

37. Ha HC, Snyder SH. Poly(ADP-ribose) polymerase is a mediator of necrotic cell death by ATP depletion. Proc Natl Acad Sci USA 1999; 96:13978-82.

38. Han D, Williams E, Cadenas E. Mitochondrial respiratory chain-dependent generation of superoxide anion and its release into the intermembrane space. Biochem J 2001; 353:411-6.

39. Harper N, Hughes M, MacFarlane M et al. Fas-associated death domain protein and caspase-8 are not recruited to the tumor necrosis factor receptor 1 signaling complex during tumor necrosis factor-induced apoptosis. J Biol Chem 2003; 278:25534-41.

40. Hasegawa J, Kamada S, Kamiike W et al. Involvement of CPP32/Yama(-like) proteases in Fas-mediated apoptosis. Cancer Res 1996; 56:1713-8.

41. Herceg Z, Wang ZQ. Failure of poly(ADP-ribose) polymerase cleavage by caspases leads to induction of necrosis and enhanced apoptosis. Mol Cell Biol 1999; 19:5124-33.

42. Hetz CA, Hunn M, Rojas P et al. Caspase-dependent initiation of apoptosis and necrosis by the Fas receptor in lymphoid cells: Onset of necrosis is associated with delayed ceramide increase. J Cell Sci 2002; 115:4671-83.

43. Heyninck K, Denecker G, De Valck D et al. Inhibition of tumor necrosis factor-induced necrotic cell death by the zinc finger protein A20. Anticancer Res 1999; 19:2863-8.

44. Higuchi Y. Chromosomal DNA fragmentation in apoptosis and necrosis induced by oxidative stress. Biochem Pharmacol 2003; 66:1527-35.

45. Hill JM, Morisawa G, Kim T et al. Identification of an expanded binding surface on the FADD death domain responsible for interaction with CD95/Fas. J Biol Chem 2003.

46. Holler N, Zaru R, Micheau O et al. Fas triggers an alternative, caspase-8-independent cell death pathway using the kinase RIP as effector molecule. Nat Immunol 2000; 1:489-95.

47. Hsu H, Huang J, Shu HB et al. TNF-dependent recruitment of the protein kinase RIP to the TNF receptor-1 signaling complex. Immunity 1996; 4:387-96.

48. Huang DC, Hahne M, Schroeter M et al. Activation of Fas by FasL induces apoptosis by a mechanism that cannot be blocked by Bcl-2 or Bcl-x(L). Proc Natl Acad Sci USA 1999; 96:14871-6.

49. Hueber AO, Bernard AM, Herincs Z et al. An essential role for membrane rafts in the initiation of Fas/CD95-triggered cell death in mouse thymocytes. EMBO Rep 2002; 3:190-6.

50. Ichas F, Jouaville LS, Mazat JP. Mitochondria are excitable organelles capable of generating and conveying electrical and calcium signals. Cell 1997; 89:1145-53.

51. Itoh N, Nagata S. A novel protein domain required for apoptosis. Mutational analysis of human Fas antigen. J Biol Chem 1993; 268:10932-7.

52. Itoh N, Yonehara S, Ishii A et al. The polypeptide encoded by the cDNA for human cell surface antigen Fas can mediate apoptosis. Cell 1991; 66:233-43.

53. Jayanthi S, Ordonez S, McCoy MT et al. Dual mechanism of Fas-induced cell death in neuroglioma cells: A role for reactive oxygen species. Brain Res Mol Brain Res 1999; 72:158-65.

54. Kalai M, Van Loo G, Vanden Berghe T et al. Tipping the balance between necrosis and apoptosis in human and murine cells treated with interferon and dsRNA. Cell Death Differ 2002; 9:981-94.

55. Kalra J, Lautner D, Massey KL et al. Oxygen free radicals induced release of lysosomal enzymes in vitro. Mol Cell Biochem 1988; 84:233-8.

56. Kawahara A, Ohsawa Y, Matsumura H et al. Caspase-independent cell killing by Fas-associated protein with death domain. J Cell Biol 1998; 143:1353-60.

57. Kawane K, Fukuyama H, Yoshida H et al. Impaired thymic development in mouse embryos deficient in apoptotic DNA degradation. Nat Immunol 2003; 4:138-44.

58. Khwaja A, Tatton L. Resistance to the cytotoxic effects of tumor necrosis factor alpha can be overcome by inhibition of a FADD/caspase-dependent signaling pathway. J Biol Chem 1999; 274:36817-23.

59. Kim JS, He L, Lemasters JJ. Mitochondrial permeability transition: A common pathway to necrosis and apoptosis. Biochem Biophys Res Commun 2003a; 304:463-70.

60. Kim JS, Qian T, Lemasters JJ. Mitochondrial permeability transition in the switch from necrotic to apoptotic cell death in ischemic rat hepatocytes. Gastroenterology 2003b; 124:494-503.

61. Kim JW, Choi EJ, Joe CO. Activation of death-inducing signaling complex (DISC) by pro-apoptotic C-terminal fragment of RIP. Oncogene 2000; 19:4491-9.

62. Kischkel FC, Hellbardt S, Behrmann I et al. Cytotoxicity-dependent APO-1 (Fas/CD95)-associated proteins form a death-inducing signaling complex (DISC) with the receptor. Embo J 1995; 14:5579-88.

63. Kohjimoto Y, Kennington L, Scheid CR et al. Role of phospholipase A2 in the cytotoxic effects of oxalate in cultured renal epithelial cells. Kidney Int 1999; 56:1432-41.

64. Korshunov SS, Skulachev VP, Starkov AA. High protonic potential actuates a mechanism of production of reactive oxygen species in mitochondria. FEBS Lett 1997; 416:15-8.

65. Kothari S, Cizeau J, McMillan-Ward E et al. BNIP3 plays a role in hypoxic cell death in human epithelial cells that is inhibited by growth factors EGF and IGF. Oncogene 2003; 22:4734-44.

66. Kowaltowski AJ, Castilho RF, Vercesi AE. Mitochondrial permeability transition and oxidative stress. FEBS Lett 2001; 495:12-5.

67. Kubasiak LA, Hernandez OM, Bishopric NH et al. Hypoxia and acidosis activate cardiac myocyte death through the Bcl-2 family protein BNIP3. Proc Natl Acad Sci USA 2002; 99:12825-30.

68. Lacronique V, Mignon A, Fabre M et al. Bcl-2 protects from lethal hepatic apoptosis induced by an anti-Fas antibody in mice. Nat Med 1996; 2:80-6.

69. Lamkanfi M, Declercq W, Depuydt B et al. The caspase family. In: Los M, Walczak H, eds. Caspases: Their Role in Cell Death and Cell Survival. Georgetown, TX: Landes Bioscience, Kluwer Academic Press, 2003.

70. Lamy L, Ticchioni M, Rouquette-Jazdanian AK et al. CD47 and the 19 kDa interacting protein-3 (BNIP3) in T cell apoptosis. J Biol Chem 2003; 278:23915-21.

71. Lardot C, Broeckaert F, Lison D et al. Exogenous catalase may potentiate oxidant-mediated lung injury in the female Sprague-Dawley rat. J Toxicol Environ Health 1996; 47:509-22.

72. Latta M, Kunstle G, Leist M et al. Metabolic depletion of ATP by fructose inversely controls CD95- and tumor necrosis factor receptor 1-mediated hepatic apoptosis. J Exp Med 2000; 191:1975-85.

73. Legembre P, Moreau P, Daburon S et al. Potentiation of Fas-mediated apoptosis by an engineered glycosylphosphatidylinositol-linked Fas. Cell Death Differ 2002; 9:329-39.

74. Leist M, Nicotera P. The shape of cell death. Biochem Biophys Res Commun 1997; 236:1-9.

75. Leist M, Single B, Naumann H et al. Nitric oxide inhibits execution of apoptosis at two distinct ATP-dependent steps upstream and downstream of mitochondrial cytochrome c release. Biochem Biophys Res Commun 1999; 258:215-21.

76. Lewis J, Devin A, Miller A et al. Disruption of hsp90 function results in degradation of the death domain kinase, receptor-interacting protein (RIP), and blockage of tumor necrosis factor-induced nuclear factor-kappaB activation. J Biol Chem 2000; 275:10519-26.

77. Li H, Zhu H, Xu CJ et al. Cleavage of BID by caspase 8 mediates the mitochondrial damage in the Fas pathway of apoptosis. Cell 1998; 94:491-501.

78. Li P, Nijhawan D, Budihardjo I et al. Cytochrome c and dATP-dependent formation of Apaf-1/caspase-9 complex initiates an apoptotic protease cascade. Cell 1997; 91:479-89.

79. Li W, Yuan X, Nordgren G et al. Induction of cell death by the lysosomotropic detergent MSDH. FEBS Lett 2000; 470:35-9.

80. Lin Y, Devin A, Rodriguez Y et al. Cleavage of the death domain kinase RIP by caspase-8 prompts TNF-induced apoptosis. Genes Dev 1999; 13:2514-26.

81. Lockshin RA, Williams CM. Programmed cell death. II. Endocrine potentiation of the breakdown of the intersegmental muscles of silkmoths. J. Insect Physiol 1964; 10:643-649.

82. Los M, Mozoluk M, Ferrari D et al. Activation and caspase-mediated inhibition of PARP: A molecular switch between fibroblast necrosis and apoptosis in death receptor signaling. Mol Biol Cell 2002; 13:978-88.

83. Los M, Van de Craen M, Penning LC et al. Requirement of an ICE/CED-3 protease for Fas/APO-1-mediated apoptosis. Nature 1995; 375:81-3.

84. Martinon F, Holler N, Richard C et al. Activation of a pro-apoptotic amplification loop through inhibition of NF-kappaB-dependent survival signals by caspase-mediated inactivation of RIP. FEBS Lett 2000; 468:134-6.

85. Mateo V, Lagneaux L, Bron D et al. CD47 ligation induces caspase-independent cell death in chronic lymphocytic leukemia. Nat Med 1999; 5:1277-84.

86. Meilhac O, Escargueil-Blanc I, Thiers JC et al. Bcl-2 alters the balance between apoptosis and necrosis, but does not prevent cell death induced by oxidized low density lipoproteins. Faseb J 1999; 13:485-94.

87. Memon SA, Moreno MB, Petrak D et al. Bcl-2 blocks glucocorticoid- but not Fas- or activation-induced apoptosis in a T cell hybridoma. J Immunol 1995; 155:4644-52.

88. Micheau O, Tschopp J. Induction of TNF receptor I-mediated apoptosis via two sequential signaling complexes. Cell 2003; 114:181-90.

89. Mohr S, Zech B, Lapetina EG et al. Inhibition of caspase-3 by S-nitrosation and oxidation caused by nitric oxide. Biochem Biophys Res Commun 1997; 238:387-91.

90. Morel Y, Barouki R. Repression of gene expression by oxidative stress. Biochem J 1999; 342(Pt 3):481-96.

91. Muller S, Scaffidi P, Degryse B et al. New EMBO members' review: The double life of HMGB1 chromatin protein: Architectural factor and extracellular signal. Embo J 2001; 20:4337-40.

92. Murakami M, Kudo I. Phospholipase A2. J Biochem (Tokyo) 2002; 131:285-92.

93. Muzio M, Chinnaiyan AM, Kischkel FC et al. FLICE, a novel FADD-homologous ICE/CED-3-like protease, is recruited to the CD95 (Fas/APO-1) death—inducing signaling complex. Cell 1996; 85:817-27.

94. Nakayama M, Ishidoh K, Kayagaki N et al. Multiple pathways of TWEAK-induced cell death. J Immunol 2002; 168:734-43.

95. Napirei M, Wulf S, Mannherz HG. Chromatin breakdown during necrosis by serum Dnase1 and the plasminogen system. Arthritis & Rheumatism 2004; In press.

96. Ollinger K, Brunk UT. Cellular injury induced by oxidative stress is mediated through lysosomal damage. Free Radic Biol Med 1995; 19:565-74.

97. Orrenius S, Zhivotovsky B, Nicotera P. Regulation of cell death: The calcium-apoptosis link. Nat Rev Mol Cell Biol 2003; 4:552-65.

98. Papoff G, Hausler P, Eramo A et al. Identification and characterization of a ligand-independent oligomerization domain in the extracellular region of the CD95 death receptor. J Biol Chem 1999; 274:38241-50.

99. Peter ME, Krammer PH. The CD95(APO-1/Fas) DISC and beyond. Cell Death Differ 2003; 10:26-35.

100. Raymond MA, Mollica L, Vigneault N et al. Blockade of the apoptotic machinery by cyclosporin A redirects cell death toward necrosis in arterial endothelial cells: Regulation by reactive oxygen species and cathepsin D. Faseb J 2003; 17:515-7.

101. Rizzuto R, Pinton P, Ferrari D et al. Calcium and apoptosis: Facts and hypotheses. Oncogene 2003; 22:8619-27.

102. Sapirstein A, Spech RA, Witzgall R et al. Cytosolic phospholipase A2 (PLA2), but not secretory PLA2, potentiates hydrogen peroxide cytotoxicity in kidney epithelial cells. J Biol Chem 1996; 271:21505-13.

103. Scaffidi C, Fulda S, Srinivasan A et al. Two CD95 (APO-1/Fas) signaling pathways. Embo J 1998; 17:1675-87.

104. Scaffidi C, Schmitz I, Zha J et al. Differential modulation of apoptosis sensitivity in CD95 type I and type II cells. J Biol Chem 1999; 274:22532-8.

105. Scaffidi P, Misteli T, Bianchi ME. Release of chromatin protein HMGB1 by necrotic cells triggers inflammation. Nature 2002; 418:191-5.

106. Schulze-Osthoff K, Beyaert R, Vandevoorde V et al. Depletion of the mitochondrial electron transport abrogates the cytotoxic and gene-inductive effects of TNF. Embo J 1993; 12:3095-104.
107. Schweichel JU, Merker HJ. The morphology of various types of cell death in prenatal tissues. Teratology 1973; 7:253-66.
108. Shinoura N, Yoshida Y, Asai A et al. Relative level of expression of Bax and Bcl-XL determines the cellular fate of apoptosis/necrosis induced by the overexpression of Bax. Oncogene 1999; 18:5703-13.
109. Siegel RM, Frederiksen JK, Zacharias DA et al. Fas preassociation required for apoptosis signaling and dominant inhibition by pathogenic mutations. Science 2000; 288:2354-7.
110. Smith CA, Farrah T, Goodwin RG. The TNF receptor superfamily of cellular and viral proteins: Activation, costimulation, and death. Cell 1994; 76:959-62.
111. Sowter HM, Ratcliffe PJ, Watson P et al. HIF-1-dependent regulation of hypoxic induction of the cell death factors BNIP3 and NIX in human tumors. Cancer Res 2001; 61:6669-73.
112. Spencer MJ, Croall DE, Tidball JG. Calpains are activated in necrotic fibers from mdx dystrophic mice. J Biol Chem 1995; 270:10909-14.
113. Spiteller G. Are lipid peroxidation processes induced by changes in the cell wall structure and how are these processes connected with diseases? Med Hypotheses 2003; 60:69-83.
114. Strasser A, Harris AW, Huang DC et al. Bcl-2 and Fas/APO-1 regulate distinct pathways to lymphocyte apoptosis. Embo J 1995; 14:6136-47.
115. Syntichaki P, Tavernarakis N. Death by necrosis. Uncontrollable catastrophe, or is there order behind the chaos? EMBO Rep 2002; 3:604-9.
116. Syntichaki P, Xu K, Driscoll M et al. Specific aspartyl and calpain proteases are required for neurodegeneration in C. elegans. Nature 2002; 419:939-44.
117. Tavernarakis N, Xu K, Driscoll M. Execution of necrotic-like cell death in caenorhabditis elegans requires cathepsin d activity. Scientific World Journal 2001; 1:139.
118. Tepper AD, Ruurs P, Wiedmer T et al. Sphingomyelin hydrolysis to ceramide during the execution phase of apoptosis results from phospholipid scrambling and alters cell-surface morphology. J Cell Biol 2000; 150:155-64.
119. Thomas LR, Stillman DJ, Thorburn A. Regulation of Fas-associated death domain interactions by the death effector domain identified by a modified reverse two-hybrid screen. J Biol Chem 2002; 277:34343-8.
120. Trauth BC, Klas C, Peters AM et al. Monoclonal antibody-mediated tumor regression by induction of apoptosis. Science 1989; 245:301-5.
121. Tsukada T Watanabe M, Yamashima T. Implications of CAD and DNase II in ischemic neuronal necrosis specific for the primate hippocampus. J Neurochem 2001; 79:1196-206.
122. Turpaev KT. Reactive oxygen species and regulation of gene expression. Biochemistry (Mosc) 2002; 67:281-92.
123. van Loo G, Saelens X, van Gurp M et al. The role of mitochondrial factors in apoptosis: A Russian roulette with more than one bullet. Cell Death Differ 2002; 9:1031-42.
124. Vande Velde C, Cizeau J, Dubik D et al. BNIP3 and genetic control of necrosis-like cell death through the mitochondrial permeability transition pore. Mol Cell Biol 2000; 20:5454-68.
125. Vanden Berghe T, Kalai M, van Loo G et al. Disruption of HSP90 function reverts tumor necrosis factor-induced necrosis to apoptosis. J Biol Chem 2003a; 278:5622-9.
126. Vanden Berghe T, van Loo G, Saelens X et al. Differential signaling to apoptotic and necrotic cell death by Fas-associated death domain protein FADD. J Biol Chem 2003b; in press.
127. Vander Heiden MG, Thompson CB. Bcl-2 proteins: Regulators of apoptosis or of mitochondrial homeostasis? Nat Cell Biol 1999; 1:E209-16.
128. Varfolomeev EE, Boldin MP, Goncharov TM et al. A potential mechanism of "cross-talk" between the p55 tumor necrosis factor receptor and Fas/APO1: Proteins binding to the death domains of the two receptors also bind to each other. J Exp Med 1996; 183:1271-5.
129. Vercammen D, Beyaert R, Denecker G et al. Inhibition of caspases increases the sensitivity of L929 cells to necrosis mediated by tumor necrosis factor. J Exp Med 1998a; 187:1477-85.
130. Vercammen D, Brouckaert G, Denecker G et al. Dual signaling of the Fas receptor: Initiation of both apoptotic and necrotic cell death pathways. J Exp Med 1998b; 188:919-30.
131. Virag L, Szabo C. The therapeutic potential of poly(ADP-ribose) polymerase inhibitors. Pharmacol Rev 2002; 54:375-429.

132. Wang KK. Calpain and caspase: Can you tell the difference? Trends Neurosci 2000; 23:20-6.

133. Wang X, Ryter SW, Dai C et al. Necrotic cell death in response to oxidant stress involves the activation of the apoptogenic caspase-8/bid pathway. J Biol Chem 2003; 278:29184-91.

134. Wang ZQ, Stingl L, Morrison C et al. PARP is important for genomic stability but dispensable in apoptosis. Genes Dev 1997; 11:2347-58.

135. Xu K, Tavernarakis N, Driscoll M. Necrotic cell death in C. elegans requires the function of calreticulin and regulators of Ca(2+) release from the endoplasmic reticulum. Neuron 2001; 31:957-71.

136. Yamashima T. Implication of cysteine proteases calpain, cathepsin and caspase in ischemic neuronal death of primates. Prog Neurobiol 2000; 62:273-95.

137. Yamashima T, Kohda Y, Tsuchiya K et al. Inhibition of ischaemic hippocampal neuronal death in primates with cathepsin B inhibitor CA-074: A novel strategy for neuroprotection based on 'calpain-cathepsin hypothesis'. Eur J Neurosci 1998; 10:1723-33.

138. Yamashima T, Saido TC, Takita M et al. Transient brain ischaemia provokes Ca2+, PIP2 and calpain responses prior to delayed neuronal death in monkeys. Eur J Neurosci 1996; 8:1932-44.

139. Yamashima T, Tonchev AB, Tsukada T et al. Sustained calpain activation associated with lysosomal rupture executes necrosis of the postischemic CA1 neurons in primates. Hippocampus 2003; 13:791-800.

140. Yonehara S, Ishii A, Yonehara M. A cell-killing monoclonal antibody (anti-Fas) to a cell surface antigen codownregulated with the receptor of tumor necrosis factor. J Exp Med 1989; 169:1747-56.

141. Yu Z, Persson HL, Eaton JW et al. Intralysosomal iron: A major determinant of oxidant-induced cell death. Free Radic Biol Med 2003; 34:1243-52.

142. Zhang J, Liu X, Scherer DC et al. Resistance to DNA fragmentation and chromatin condensation in mice lacking the DNA fragmentation factor 45. Proc Natl Acad Sci USA 1998; 95:12480-5.

143. Zhao M, Antunes F, Eaton JW et al. Lysosomal enzymes promote mitochondrial oxidant production, cytochrome c release and apoptosis. Eur J Biochem 2003; 270:3778-86.

144. Zhu LP, Yu XD, Ling S et al. Mitochondrial Ca(2+)homeostasis in the regulation of apoptotic and necrotic cell deaths. Cell Calcium 2000; 28:107-17.

CHAPTER 6

Fas—More Than an Apoptosis Inducer

Harald Wajant

Abstract

Fas (Apo-1 or CD95) and its corresponding ligand FasL (CD95L) are representative members of the TNF receptor and TNF ligand family that have been implicated in a variety of apoptotic processes, involved in T-cell induced cytotoxicity, activation-induced cell death, immune privilege, tumor surveillance and angiogenesis. Although, studies on the FasL/Fas system mainly focused on its pro-apoptotic role, a couple of additional apoptosis-independent functions of Fas have been reported, including induction of proliferation in T-cells and fibroblasts, hepatocyte regeneration, chemokine production, DC maturation and neurite outgrowth. While the apoptotic signaling capacities of FasL and Fas were intensively studied and well understood, the molecular mechanisms of nonapoptotic Fas signaling are ill defined yet. This chapter will review our current understanding of nonapoptotic FasL/Fas functions and in particular address how the balance between apoptotic and nonapoptotic Fas signaling is regulated.

Introduction

The structure and transcriptional regulation of FasL and Fas have been described in detail in chapter 1 and 9. Like other proteins of the TNF ligand family, FasL is a trimeric type II membrane protein, which can be converted into a soluble trimeric protein by proteolysis or alternative splicing.[1,2] Although membrane as well as soluble FasL bind to Fas, the capacity of membrane FasL to induce Fas signaling is by several orders higher as compared to soluble FasL.[3-5] Fas is a type I transmembrane protein and has three "cysteine-rich domains" in its extracellular domain that are typical for members of the TNF receptor superfamily.[6] In its intracellular domain Fas contains a death domain (DD) which is necessary for apoptosis signaling.[7] A death domain is not only part of Fas and other death receptors, but it is also found in cytoplasmic adaptor proteins involved in apoptosis signaling.[8] Death domains function as protein-protein-interaction modules allowing homo- and/or heteromerization of death domain-containing proteins. In nonstimulated cells Fas preaggregates in signaling incompetent complexes by virtue of its amino-terminal preligand-binding assembly domain.[9-11] Binding of membrane FasL triggers the reorganization of these complexes into signaling competent ligand-receptor aggregates. These aggregates are capable to form supramolecular clusters and interact with cytoplasmic signaling molecules. In sensitive cells this "active" Fas complex induces apoptosis and has been therefore named DISC (death inducing signaling complex).[12] Aside FasL and Fas the DISC contains the death domain-containing adaptor protein FADD (Fas-associated death domain protein) and procaspase-8 as well as processing intermediates of the

Fas Signaling, edited by Harald Wajant. ©2006 Landes Bioscience and Springer Science+Business Media.

latter.[13] While FADD binds directly to the DD of Fas by its own C-terminal DD, procaspase-8 is indirectly recruited to Fas via FADD, which interacts via its N-terminal death effector domain with the corresponding structure in procaspase-8.[13-15] This FADD-mediated recruitment into the DISC allows the transient formation of enzymatically active procaspase-8 dimers that convert by autoproteolytic processing to mature active caspases-8 heterotetramers.[16,17] Mature caspase-8 is released from the DISC and, dependent on the cell type, can trigger the execution phase of apoptosis by two pathways. First, mature caspase-8 can activate effector caspase, in particular caspase-3, by proteolytic processing. In type I cells this is sufficient for a robust activation of the effector steps of apoptosis.[18,19] However, in some cells, called type II cells, this caspase cascade is not or only inefficiently activated due to too low amounts of mature caspase-8 and / or the action of caspase inhibitory molecules. In these cases apoptosis induction requires an amplification mechanism which is initially triggered by caspase-8-mediated cleavage of Bid, a BH3-only protein of the Bcl-2 family. Cleavage of Bid results in truncated Bid fragment (tBid), which translocates to the mitochondria where it binds and allosterically activates Bak (and possibly Bax).[18,19] The intramembranous oligomerization of the latter allows the release of mitochondrial apoptogenic proteins, including cytochrome c, the second mitochondria-derived activator of caspase / direct IAP binding protein with low PI (SMAC/Diablo) and HtrA2/Omi.[20] Cytosolic cytochrome c, together with ATP and the scaffold protein apoptosis promoting factor-1 (Apaf-1), binds and activates caspase-9 which as caspase-8 stimulates caspase-3 activation.[21] Smac/Diablo and HtrA2/Omi support further caspase-3 activation by initiator caspases, by blocking the caspase-inhibitory members of the IAP protein family.[20] Cells that crucially utilize the mitochondria-dependent pathway to convert Fas and caspase-8 activation into apoptosis can therefore be rescued by ectopic overexpression of Bcl-2 or Bcl-XL. In contrast, in type I cells where caspase-3 activation is a direct consequence of Fas and caspase-8 activation, expression of these anti-apoptotic proteins has no major effect.[18,19] In some cells, Fas causes an alternative caspase-independent form of cell death called necrosis.[22-24] Although FADD and the Fas-interacting serine/threonine kinase RIP, have been recently identified as critical mediators of Fas-induced necrosis, the linkage of these receptor associated signaling intermediates to the production of reactive oxygen species and other effector mechanisms of necrosis remains unknown.[22] Cell death induction by Fas is regulated by a variety of intracellular and extracellular factors which enable cells to respond highly flexible towards Fas stimulation. According to the particular step in Fas signaling, which is targeted by regulatory molecules, regulation is Fas-specific, common to death receptors or of relevance for the apoptotic core machinery of the cell.

Fas-Induced NFκB Activation

NFκB and IκB Proteins

The term nuclear factor of kappa light polypeptide gene enhancer in B-cells (NFκB) refers to a group of dimeric transcription factors composed of homo- and/or heterodimers of proteins of the phylogenetically conserved NFκB/Rel family. The structural hallmark of these proteins is the Rel homology domain (RHD) which is involved in dimerization, nuclear localization, DNA binding and interaction with the inhibitory counterparts of the NFκB proteins—the IκB proteins. Five mammalian NFκB proteins are known, p65/RelA, RelB, cRel, NFκB1/p50 and NFκB2/p52, whereby the latter two are produced by constitutive and / or inducible proteolysis of the precursor proteins p105 (p50) and p100 (p52).[25,26] RelA/p65, cRel and RelB but not p50 and p52 contain a carboxy-terminal transactivation domain (Fig. 1). Homodimers of the latter two act therefore as transcriptional repressors while heteromers of p50 or p52 with RelA, cRel or RelB are transcriptionally active. In general, the various NFκB dimers interact with high affinity with most κB DNA binding sites. The relative strength of interaction between various distinct NFκB dimers and distinct promoters is fine tuned by the expression levels of NFκB proteins,

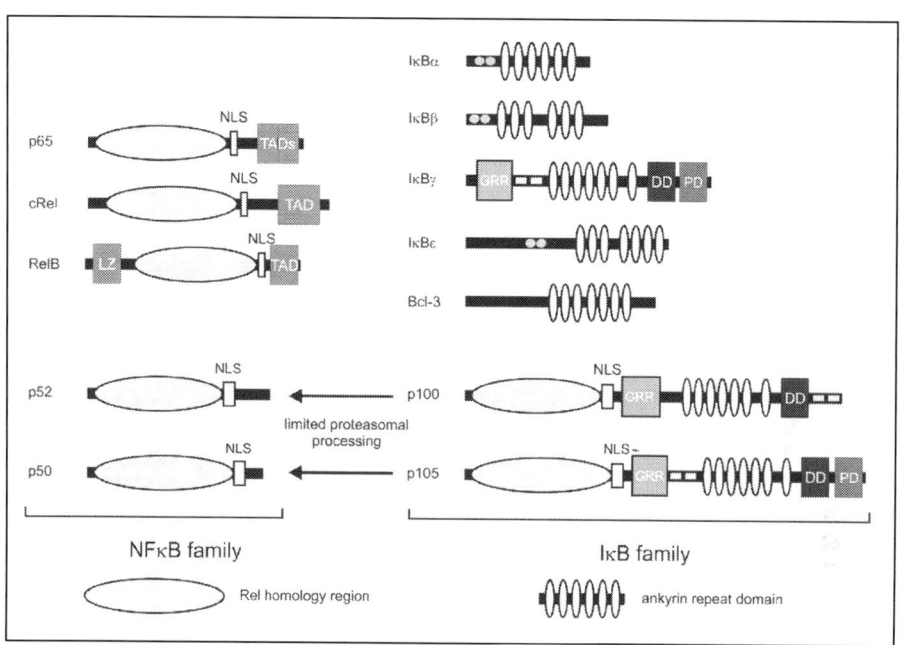

Figure 1. Scheme of mammalian NFκB and IκB proteins. The Rel homology region consisting of a DNA binding domain, a dimerization domain and a nuclear localization sequence (NLS) is the structural hallmark of the NFκB transcription factors. The ankyrin repeat domain is the characteristic feature of the IκBα proteins. Other structural features contained in NFκB or IκBα proteins are leucine zippers (LZ), transactivation domains (TAD), glycine-rich regions (GRR), death domains (DD) and PEST domains (PD). The serines of the signal response domain of IκBα, IκBβ and IκBε are depicted as gray ellipses. Regulatory phosphorylation sites in p105 and IκBγ are shown as gray boxes. IκBγ is identical to the C-terminal part of p105.

their affinities to these promoters, the presence of NFκB interacting proteins and posttranslational modifications.[25,26] In nonstimulated cells NFκB dimers are sequestered in the cytoplasm by forming a ternary complex with IκB proteins. These proteins are characterized by six or seven ankyrin repeats mediating interaction with the RHD of NFκB proteins. The IκB family comprises in mammals IκBα, IκBβ, IκBγ, IκBε and Bcl-3, as well as 105 and p100 the aforementioned precursor proteins of the NFκB proteins p50 and p52 (Fig. 1).[25,26] The IκB protein interacts in a ternary NFκB-IκB complex asymmetrically with the NFκB homo- or heterodimers thereby masking differentially the nuclear localization sequences of the NFκB subunits.[27,28] The stability and specificity of the ternary NFκB-IκB complex as a whole is mainly dependent on the interaction of the IκB protein with one of the subunits of the NFκB dimer in the complex.

IκBα, IκBβ and IκBε have an amino-terminal regulatory domain, containing a pair of conserved serine residues that undergo phosphorylation upon triggering of the NFκB pathway.[29-31] Such modified IκB proteins are recognized by the Skp1-Cullin-F-box protein beta-transducin repeat-containing protein (SCF^{βTRCP}) ubiquitin ligase complex, which ubiquitinates two lysine residues adjacent to the phospho-serines of the regulatory domain.[32] Ubiquitinated IκB proteins in turn are degraded by the 26S proteasome.[33,34] p105 and p100 dimerize with a variety of NFκB proteins (p105, refs. 35-37), most importantly with RelB (p100, ref. 38), to inactive cytoplasmic complexes resembling ternary IκB-NFκB complexes. NFκB inducing signals trigger limited processing or degradation of such precursor protein containing complexes leading to the release

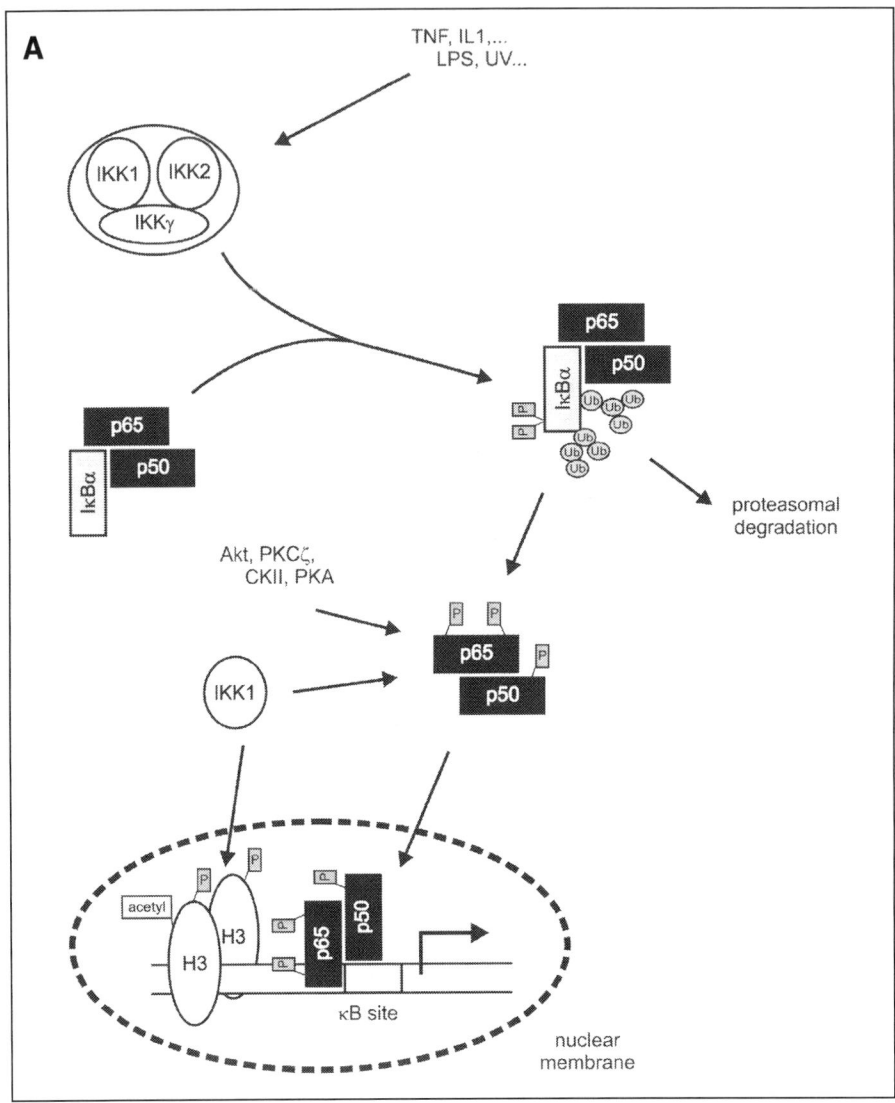

Figure 2. Classical and nonclassical activation. The canonical or classical NFκB pathway (A) is for example triggered by cytokines or microbe associated products and dependent on IKK1, IKK2 and IKKγ and subsequent phosphorylation and degradation of IκB. Phoshorylation of the NFκB proteins themselves (most importantly p50 and p65) and histone 3 modulates NFκB activity. Figure continued on next page.

of NFκB dimers containing mature p50 or 52. Limited processing requires a glycine-rich region between the N-terminal NFκB domain and the C-terminal IκB domain and prevents the entry of the N-terminal NFκB fragment of p100 and p105 into the proteasome.[39-41]

Classical and Nonclassical NFκB Activation

The signal-induced degradation / processing of IκB proteins and p105 on the one hand side and p100 on the other side, thus the activation of the corresponding cytoplasmic NFκB

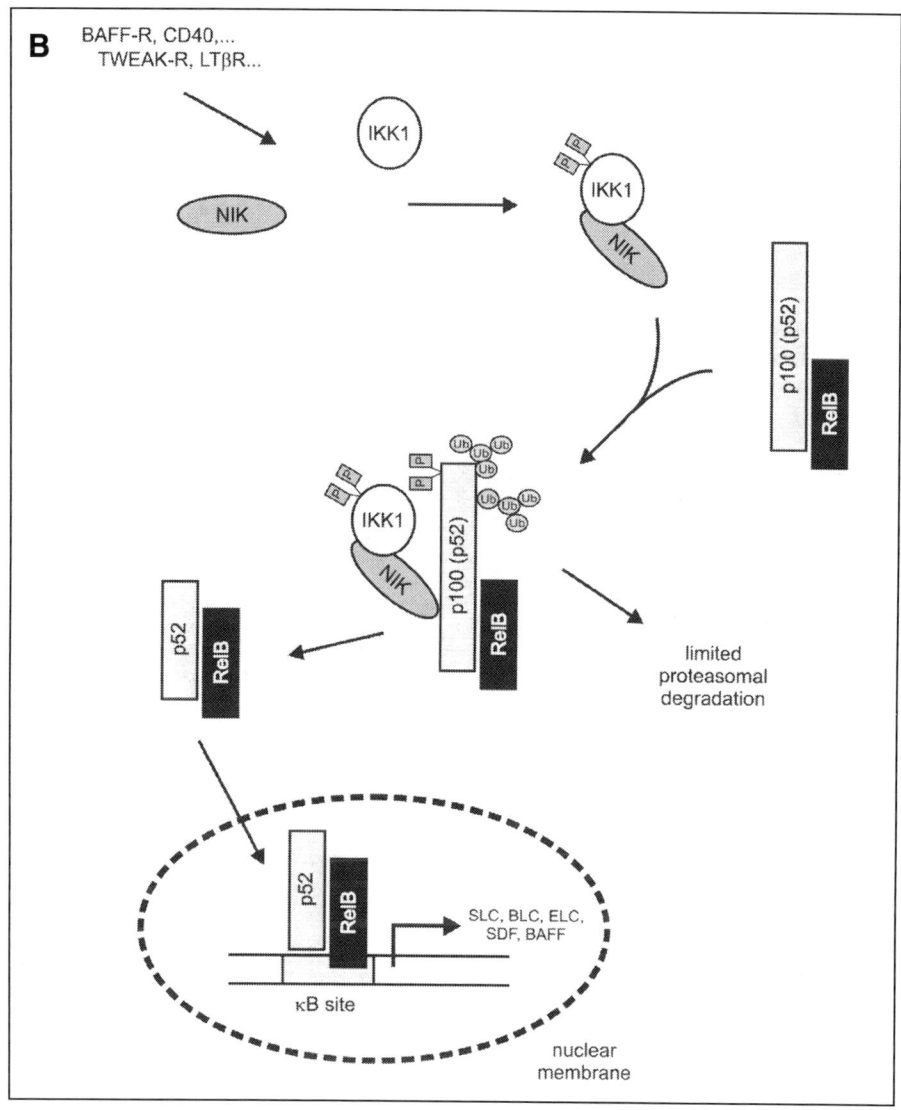

Figure 2. Continued. The nonclassical NFκB pathway (B) is triggered by a selected range of TNF receptors, is dependent on IKK1 and NIK and leads to phoshorylation-dependent processing of p100.

complexes, is differentially regulated. The so called canonical or classical NFκB triggers the degradation of IκBs and p105, while the nonclassical pathway of NFκB activation regulates the processing of p100 (Fig. 2). The majority of NFκB effects are mediated by the classical pathway, whereas the nonclassical pathway is of special relevance for B-cell development.

The IKK Complex and the Classical NFκB Pathway

Phosphorylation of a serine pair which marks IκB proteins and p105 for degradation is triggered by the IκB kinase (IKK) complex. This complex consists of homo- or heterodimers of the related serine kinases IKK1 and IKK2 which are bound to the regulatory subunit NEMO, also called IKKγ, IKKAP1 or FIP3.[42] Just recently a 110 ka protein which contains abundantly glutamic acid (E), leucine (L), lysine (K) and serine (S), named ELKS, has been identified as a further component of the IKK complex. There is evidence that ELKS is involved in the recruitment of IκB proteins to the IKK complex.[43] The chaperones Hsp90 and Cdc37 have also been recognized as part of the IKK complex but it is unclear whether these proteins are only relevant for folding of the IKK kinases or whether they are necessary for the activity of the complex itself.[44] A variety of NFκB activating stimuli channels at the IKK complex (Fig. 2). According to its central position in NFκB signaling, the IKK complex interacts with a considerable number of kinases either directly or indirectly via IKK-interacting adaptor proteins, e.g., TNF receptor associated factor proteins.[42,45] These kinases ensure two crucial events relevant for NFκB activation. Firstly, they stimulate the kinase activity of the IKK1 and IKK2 by phosphorylation of their activation loops. Secondly, they regulate transcriptional activity of NFκB by phosphorylation of NFκB subunits.[42] Activation of the IKK complex is mainly mediated by mitogen-activated protein kinase kinase kinases (MAP3Ks or MEKKs), especially MEKK1-3 and TAK1.[46-49] Phosphorylation of NFκB proteins by IKK complex associated kinases is rather complicated and relies on a diverse set of kinases, including protein kinase B/Akt (refs. 50,51), atypical protein kinase Cs (PKCs; refs. 52,53), the catalytic subunit of protein kinase A (PKA; ref. 54), casein kinase II (ref. 55) and IKK2 (ref. 56) itself.

In accordance with an essential role of the IKK complex in activation of cytoplasmic NFκB complexes, which contain IκB proteins or p105, degradation of the latter proteins is completely blocked in NEMO deficient cells.[57-62] With respect to IKK1 and IKK2 the situation is more complex. While in vitro assays revealed only minor functional differences between IKK1 and IKK2, analyses of knock out mice demonstrated only some overlapping but mainly nonoverlapping functions. For example, TNF-induced p105 degradation is normal in IKK1 and IKK2 deficient mouse embryonic fibroblasts (MEFs) suggesting that IKK1 and IKK2 act redundantly in this respect.[60] In contrast, TNF-induced IκBα degradation is strongly reduced in the absence of IKK2 but barely affected in IKK1 deficient cells.[63-66] However, TNF-induced transcription of NFκB target genes is significantly inhibited both in IKK1 and IKK2 deficient cells.[63-67] Thus, both IKKs crucially contribute to signal-induced transcription of NFκB target genes by at least partially distinct mechanisms. In accordance with different roles of IKK1 and IKK2 in NFκB signaling, it has been recently shown that IKK1, but not IKK2 and NEMO, interacts with NFκB regulated promoters, stimulates histone acetyltransferase activity of CBP by phosphorylation of histone H3 and enhances in this way transcription of corresponding genes (Fig. 2).[68,69]

The Nonclassical NFκB Pathway

The nonclassical NFκB pathway induces p100 processing of p100/RelB complexes and had been identified in the recent years (Fig. 2). It is activated in B-cells and stromal nonlymphoid cells upon engagement of LTβR, BAFF-R, CD40 or TWEAK-R.[70-74] All these receptors belong to the non death domain containing subgroup of the TNF receptor family and act in B-cell maturation and the development of secondary lymphoid organs. Limited proteolysis of p100 is triggered by IKK1-mediated phosphorylation of two serines in its far C-terminal end.[75,76] Activation of IKK1 in turn is mediated by the MAP3kinase NFκB-inducing kinase (NIK).[77] IKK1 acts in the nonclassical pathway independent from IKK2 and NEMO. Phosphorylation of p100 leads to its ubiquitination by the SCF$^{\beta TRCP}$ ubiquitin ligase complex and subsequent processing by the proteasome.[78]

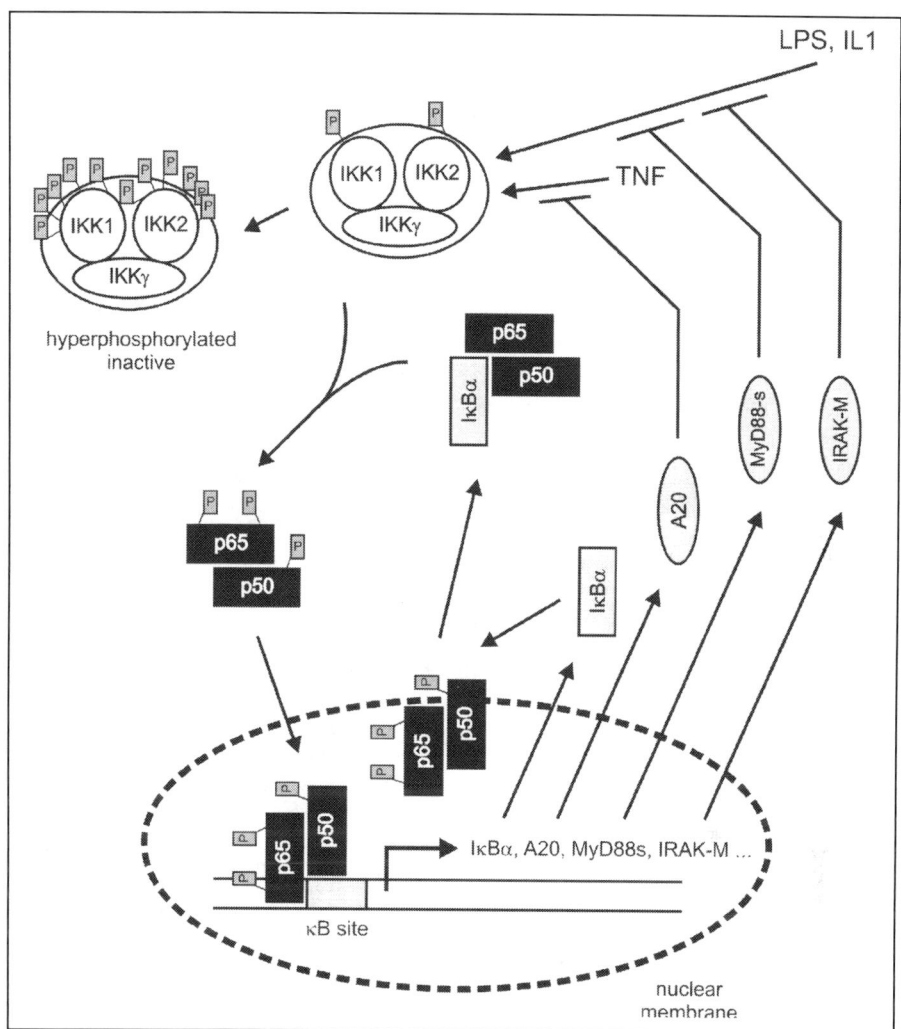

Figure 3. Multiple mechanisms are involved in the termination of NFκB activation. Activation of the NFκB pathway, thus prolonged activation of an inflammatory response, is terminated by IKK2 autophosphorylation at a serine cluster in the C-terminal part of the molecule leading to decreased IKK activity and by induction of inhibitory proteins acting at different steps of NFκB signaling.

Termination of NFκB Signaling

As an aberrant expression of NFκB target genes can have severe consequences for health, termination of NFκB signaling is therefore as important as its initiation. Accordingly, termination of NFκB is under the control of a redundant network of inhibitory feed-back loops acting at different stages of NFκB activation (Fig. 3). Dependent on the targeted step in NFκB signaling the various termination mechanisms can either globally down-regulate NFκB signaling or inhibit one or a group of related NFκB inducing stimuli. Two global mechanisms that down-regulate NFκB signaling are inactivation of the kinases of the IKK complex and induction of IκBα. Stimulatory phosphorylation of the activation loops of IKK1 and IKK2 seems to be facilitated by intramolecular interaction of a serine cluster preceding the C-terminal helix-loop-helix domain and the N-terminal kinase domain of the molecule.[79] Disturbance of this intramolecular interaction by autophosphorylation of the serine cluster might then down-regulate IKK activity. IκBα is a bona fide NFκB target gene.[80] Thus, it is rapidly resynthesized in the cytoplasm upon IKK complex-induced degradation and translocates to the nucleus. Here, it terminates transcription of NFκB target genes by interaction with transcriptionally active p50-RelA complexes leading to their release from the DNA.[80-83] Thus, autocatalytical inhibition of IKKs and NFκB-mediated upregulation of IκBα account together for a transient NFκB response. However, durable NFκB activation is also possible and is facilitated by IκBβ. Unphosphorylated, but not phosphorylated IκBβ competes with IκBα for formation of ternary complexes with DNA-bound NFκB, however, without disturbing the DNA-NFκB interaction. IκBβ thereby prevents the IκBα-mediated release of transcriptional active NFκB complexes from the promoter.[84,85] Examples of NFκB target genes that block a subgroup of NFκB inducers are A20, IL1 receptor-associated kinase-M (IRAK-M), and the short splice form of the myeloid differentiation primary response gene 88 (MyD88-s). A20 selectively blocks TNF-induced NFκB activation by inhibition of recruitment of the signaling intermediates TRADD and RIP to TNF-R1 and by triggering proteasomal degradation of TNF-R1 bound RIP.[86,87] IRAK-M and MyD88-s interfere with NFκB signaling by Toll-like receptors and the interleukin-1 receptor.[88,89]

Fas-Induced NFκB Activation and Its Relationship to Apoptosis Induction

In recent years it became evident that NFκB activation and apoptosis induction, especially by death receptors, are intimately connected by reciprocal inhibitory feed-back loops (Fig. 4). Thus, NFκB activation interferes with apoptosis induction and vice versa. The anti-apoptotic role of NFκB was first revealed in RelA deficient mice. Loss of RelA expression causes embryonic lethality at 15-16 days of gestation by TNF-induced liver cell apoptosis.[90,91] Afterwards NFκB signaling has also been shown to protect against ionizing radiation, chemotherapeutic drugs and p53-induced apoptosis.[92,93] In accordance with the broad anti-apoptotic action of NFκB a variety of apoptosis inhibiting NFκB target genes has been identified, including cellular inhibitor of apoptosis-1 (cIAP1, ref. 94), cIAP2 (ref. 95), x-linked IAP (xIAP, ref. 96), TNF-receptor associated factor-1 (TRAF1; refs. 94,97), FLIP (cFLIP; refs. 98,99), SCC-22 (ref. 100), Bcl2-related protein A1 (Bfl1/A1; refs. 101-103), Bcl-xL (refs. 103,104) and TRAIL-R3 (ref. 105). The inhibition of NFκB signaling under apoptotic conditions relies on activation of caspases. This is due to caspase-mediated cleavage of various signaling intermediates involved in NFκB activation, e.g., RIP (ref. 106,107), TRAF1 (ref. 108), IκBα (ref. 109), IKK2 (ref. 110), Akt (ref. 111), NIK (ref. 112), HPK (ref. 113) and p50 (ref. 114) and RelA (refs. 114,115). The inhibitory effect of caspase-mediated cleavage of these proteins is twofold. Firstly, the amount of available signaling intermediates necessary for NFκB signaling is reduced. Secondly, cleavage occurs side specific in a way that dominant-negative acting fragments of the noncleaved counterparts are generated.

Molecular Mechanisms of Fas-Induced NFκB Activation

For analysis of Fas-induced NFκB activation it has to be taken into consideration that this response is masked or attenuated by concomitantly induced apoptosis. In fact, it has been observed that anti-apoptotic proteins and caspase inhibitors do not only block apoptosis induction by Fas, but also potently enhance the capacity of this molecule to activate NFκB. Several studies in the recent years have shown that Fas signals NFκB activation. At the first view these data prompt a model where the FLIP-L or FLIP-S arrested Fas associated death inducing signaling complex stimulates NFκB signaling instead of apoptosis induction. Thus, incorporation of FLIP-L or FLIP-S would serve in Fas signaling as a switch between apoptosis

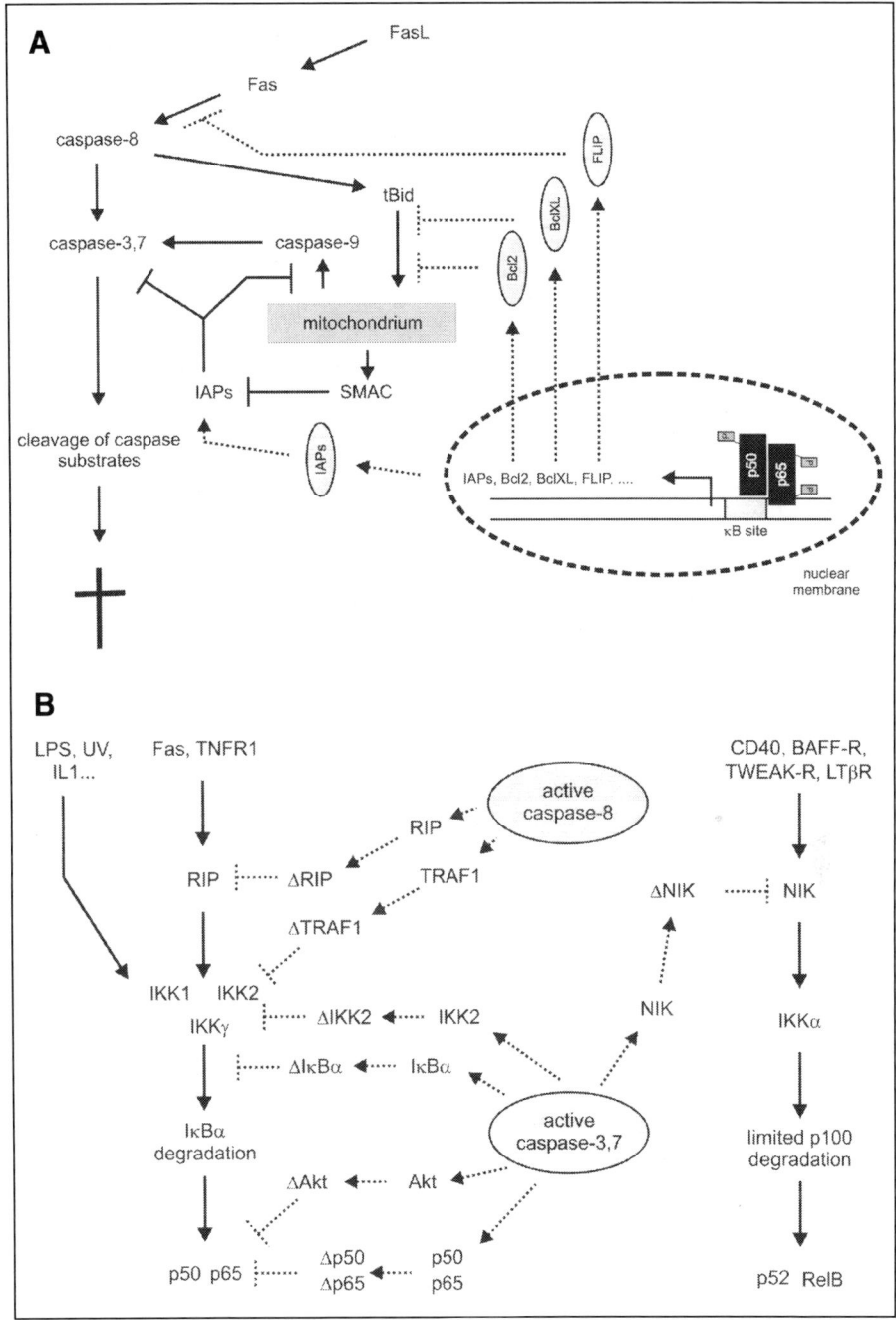

Figure 4. NFκB activation and apoptosis induction by Fas interfere with each other. NFκB-induced proteins (grey ellipses) block apoptosis-induction and caspase activation by Fas at various steps (A). Active caspases cleave components of the NFκB signaling pathway generating fragments acting in a dominant-negative fashion to their noncleaved counterparts (B).

and NFκB signaling. However, a closer look on the experimental details of the cited studies reveals some contradictions arguing against this obvious model. While NFκB activation triggered by overexpression of caspase-8 or stimulation of Fas is independent from caspase-8 activity and rather enhanced by caspase inhibitors, FADD-, FLIP$_L$- and FLIP$_S$-induced NFκB activation occurs in a way dependent on caspase-8 activity.[112,116-119] NFκB activation by these molecules is therefore inhibited by proteins (CrmA, p35) or peptides (BD-fmk, z-VAD-fmk) that inhibit caspase-8 activation.[112,118] Despite the opposing dependencies on caspase activity of FADD- and caspase-8 induced NFκB activation, we found in cells deficient for the expression of these proteins that both molecules are necessary for FasL-induced upregulation of the bona fide NFκB target gene IκBα.[119] Thus, both Fas-induced NFκB activation as well as Fas-induced apoptosis needs caspase-8, but only the latter is dependent on its catalytic properties suggesting that caspase-8 has also a scaffold function in context of the Fas signaling complex (Fig. 5). Together these data suggest an amended model where FADD and caspase-8 are involved in caspase activity-independent NFκB activation in context of Fas signaling but also in a second caspase activity-dependent, yet poorly defined mode of NFκB activation. The complexity in the relationship of FADD, Fas signaling and NFκB activation was increased again in recent studies showing (i) FADD sequesters MyD88, which has a crucial role in IL1R and TLR4 signaling and (ii) Fas-FasL interactions relieves the inhibitory FADD-MyD88 interaction.[120,121] Remarkably, in studies elucidating the IL1R/TLR4-MyD88-FADD crosstalk transient overexpression of FADD in the RAW macrophage cell line or in HEMEC cells induced no NFκB activation pointing to a cell-type specific capacity of FADD to trigger the NFκB pathway.[120] The relevance of the FADD-MyD88 crosstalk for the pro-inflammatory properties of Fas is discussed in detail under 6.5.

A further molecule that has been implicated in Fas-mediated NFκB signaling is the death domain-containing serine/threonine kinase RIP. Originally, RIP has been identified due to its association with Fas, but afterwards it has been predominantly recognized as a mediator of TNFR1-mediated NFκB activation and TRAIL-R1-induced JNK.[122-125] An additional role of RIP in Fas-induced NFκB activation can be deduced from recent studies showing lack of Fas-mediated IκBα upregulation in RIP deficient cells and inhibition of FADD and Fas-induced NFκB activation by a dominant-negative deletion mutant of RIP.[112,119] In the latter report dominant-negative RIP did not interfere with NFκB activation upon FLIP overexpression, again arguing for an involvement of caspase-8 and its interactor FADD in a Fas- and RIP-independent mode of NFκB activation. RIP has also been implicated in Fas-induced necrosis, which can be observed in some cell types. As necrosis is independent from NFκB activation, Fas-induced NFκB and necrosis signaling have to bifurcate at the level or downstream of RIP. Noteworthy, it has been shown that RIP is cleaved by caspase-8 upon Fas stimulation generating a NFκB-inhibitory dominant-negative RIP fragment.[106,107] Thus, the above mentioned enhanced NFκB activaton by Fas in the presence of caspase inhibitors is possibly in part due to an inhibition of RIP cleavage. As Fas-induced necrosis is dependent on intact RIP this finding also explains the observation that caspase inhibitors unmask or enhance the induction of this alternative form of cell death upon Fas stimulation.

FLIP in Fas-Induced NFκB Activation

It is a long known observation that in many cells efficient and rapid death receptor-induced apoptosis requires sensitization by metabolic inhibitors, e.g., cycloheximide or actionmycin D. It is commonly accepted that this reflects the need of down-regulation of one or more short-lived anti-apoptotic proteins that normally prevent apoptosis induction. There is evidence that in such cells Fas-mediated NFκB activation is also enhanced upon Fas stimulation but only when cooccurring apoptosis induction is blocked for example by caspase inhibitors or in type II cells by blocking the mitochondrial amplification loop. It appears therefore possible that NFκB

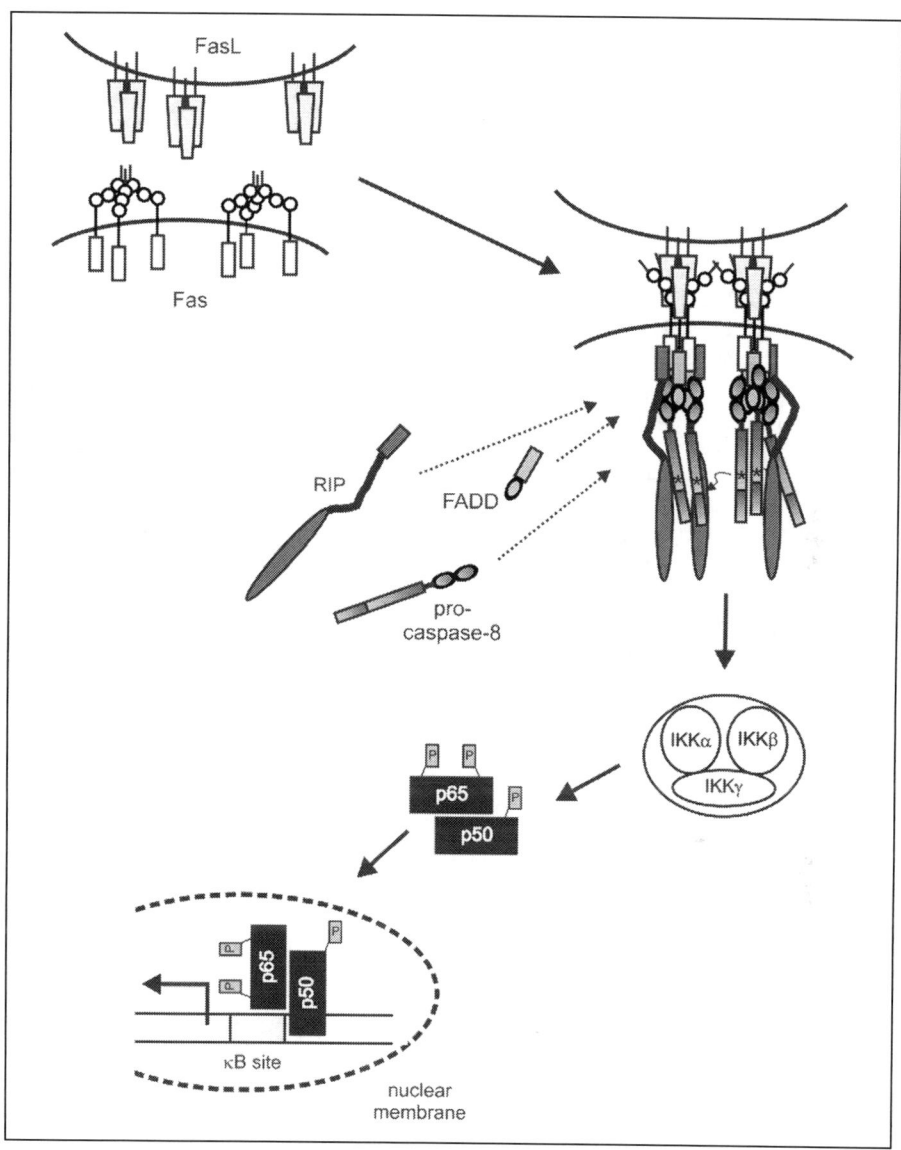

Figure 5. Fas induces NFκB activation through FADD, Caspase-8 and RIP dependent pathway which is inhibited by FLIP.

activation and apoptosis induction by Fas can be blocked by a CHX-sensitive and therefore short-lived inhibitory protein. As both responses bifurcate at the level of the Fas signaling complex, this putative inhibitor should act in a receptor proximal position. The FLIP isoforms that target the Fas-FADD-caspase-8 complex are obvious candidates for such an inhibitory factor. In fact, both FLIP-L and FLIP-S show a high turn over due to proteasomal degradation and their expression drops rapidly in the presence of inhibitors of protein synthesis.[98,116,119,126,127] Moreover, stable ectopic expression of "physiological" amounts of FLIP-L is sufficient to block FasL-induced NFκB signaling in Jurkat cells and overexpression of FLIP-L and FLIP-S, despite activating NFκB to some extend per se, inhibits Fas-induced upregulation of NFκB target genes. In further accordance with a NFκB inhibitory role of FLIP in Fas signaling, small interfering RNA oligonucleotides, either down-regulating FLIP-L and FLIP-S or only FLPI-L, not only sensitize cells for Fas-induced apoptosis but in the presence of caspase inhibitors also for NFκB activation.[119] There is evidence that in context of the Fas signaling complex procaspase-8 and FLIP-L can form an enzymatically active heteromer.[128] It is therefore tempting to speculate that cleavage of RIP and generation of the NFκB inhibitory RIP fragment take place in the FLIP-L arrested DISC. Again this idea is consistent with a FLIP-L dependent inhibition of Fas-induced NFκB signaling (Fig. 5). Inhibition of Fas-induced NFκB activation by FLIP would also explain the different modes of NFκB activation described for the death receptor TNFR1 on the one hand side and Fas on the other side. Although TNFR1, similar to Fas, signals apoptosis via a FLIP-regulated FADD- and caspase-8 dependent pathway, TNFR1-induced NFκB activation is not blocked by FLIP overexpression and occurs in cells where apoptosis induction by death receptors is dependent on sensitization with CHX.[116] Moreover, TNFR1 has been predominantly recognized in vivo as a proinflammatory molecule and proapoptotic TNFR1 signaling in vivo occurs only when concomitantly NFκB activation is blocked.[129] In agreement with these differences, TNFR1-mediated apoptosis induction and NFκB activation bifurcated upstream of the FADD-caspase-8 complex.[129]

As briefly discussed above there is evidence, mainly deduced from transient overexpression studies that FLIP acts, maybe in concert with caspase-8 and FADD, outside of the Fas signaling pathways in a poorly defined pathway leading to NFκB activation. FLIP effects not related to Fas signaling could be based on its interaction with NFκB1/p105 and p38MAPK.[130,131] Remarkably, FLIP seems to fulfil different functions in Fas signaling and in this putative second pathway. While FLIP acts as negative regulator in Fas-induced NFκB activation, it exerts NFκB inducing capacities in Fas-independent pathways. Contradictory effects of FLIP have also been found in T-cells. While TCR-induced ERK activation and IL2 production were enhanced by ectopic expression of FLIP-L in Jurkat T-cells, these responses were attenuated by FLIP in DO11.10 T-cells.[132] Analyses of primary T-cells derived from FLIP-transgenic mice revealed a similar complex regulatory role of FLIP-L. One study reported that T-cell proliferation induced by suboptimal concentrations of anti-CD3 was enhanced by FLIP, whereas a second report found that FLIP suppressed anti-CD3 induced T-cell activation over a wide range of concentrations.[133,134] Especially this second study and a recent report found that TPA- and anti-TCR/anti-CD28 induced NFκB activation was inhibited in FLIP transgenic T-cells.[134,135] As the effects of FLIP expression on Fas-induced NFκB activation have not been analyzed in these reports, future studies have to clarify whether the ambiguous effects of FLIP expression in T-cells are related to modulation of Fas-mediated NFκB activation. A Fas-independent capability to activate NFκB has also been described for the viral FLIP homologue HHV8 vFLIP.[136-139] Similar to FLIP-S this molecule mainly consists of two death effector domains. HHV8 vFLIP interacts with the IKK complex and like FLIP-S activates NFκB independent of RIP.[140] It will be interesting to learn whether FLIP-L or FLIP-S can interact with the IKK complex, too.

FasL-induced NFκB activation takes place in the absence of FLIP when apoptosis induction is blocked downstream of the Fas signaling complex. In the cited studies such circumstances were experimentally achieved by use of caspase inhibitory peptides or Bcl2 overexpression. In vivo situations allowing Fas-mediated NFκB activation may occur under pathophysiological conditions where Fas-mediated caspase-8 activation is compromised, e.g., during virus infection by specific viral genes like CrmA or in individuals expressing no or mutant caspase-8.[116,141-144] While a variety of in vitro studies has demonstrated the capability of Bcl2 to inhibit Fas-induced apoptosis, most analyses using Bcl2 transgenic or Bcl2 knockout mice argue against a major role of Bcl2 as an in vivo regulator of Fas signaling. As mice deficient for components of the mitochondrial amplification loop display consistently an enhanced susceptibility for Fas-induced liver apoptosis, this does not necessarily challenges the type I - type II cell concept but rather suggest that other proteins, distinct from Bcl2, regulate the contribution of the mitochondrial amplification loop to Fas-induced cell death in vivo. Importantly, it has to be taken into consideration that modulation of the mitochondrial branch of Fas-induced apoptosis by anti-apoptotic members of the Bcl2 family does not completely protect cells, but it rather delays apoptosis and allows for a limited time activation of the NFκB pathway. Transient protection by Bcl2 or Bcl-XL might be mechanistically related to older findings that these molecules can be inactivated by caspases. Delayed apoptosis-induction could be of special relevance for the pro-inflammatory effects of Fas activation, found in several models of Fas-dependent tumor rejection.

Fas and JNK Activation

General Aspects of the cJun N-Terminal Kinase Pathway

Mechanisms of JNK Activation

Mitogen-activated protein kinase (MAPK) cascades are phylogenetically conserved tripartite signaling modules of hierarchically acting kinases. Each of these pathways is named after the MAP kinase (MAPK) at the end of each cascade that finally phosphorylates transcription factors. MAP kinases are activated by dual specificity MAPK kinases (MAPKK) which in turn can be activated by a serine/threonine MAPKK kinase (MAPKK).[145,146] There is evidence that the signaling specificity of the various MAPK cascades is facilitated by formation of protein complexes. Aside interactions of the kinases of a given MAPK module, complex formation could also be dependent on interactions with scaffolding proteins and / or upstream activators. In mammalian cells three MAPK cascades are known (i) the extracellular signal-regulated kinases pathway (ii) the p38 kinases pathway and (iii) the cJun N-terminal kinase (JNK) pathway. Like other MAPK cascades, the latter is triggered by a variety of signals e.g., growth factors, cytokines and UV irradiation.[145,146] Activated JNK phosphorylates cJun, ATF2, JunB and JunD and thereby increases the transcription activity of AP-1 proteins containing these subunits. There are ten different mammalian JNK isoforms generated by alternative splicing from two rather ubiquitously expressed JNK genes and a third JNK gene which is predominantly transcribed in brain, heart and testis. Studies of mice deficient in one or more JNK genes showed that there is considerable redundancy between these molecules, beside nonoverlapping functions. Two MAPKKs are involved in the phosphorylation and activation of the JNK isoforms: MKK4 and MKK7. In principle, MKK4 and MKK7 phosphorylate JNKs on both threonine and tyrosine but there is clear preference of MKK4 for phosphorylation of tyrosine and of MKK7 to phosphorylate JNK on threonine. A variety of MAPKKKs, including MEKK1-4, ASK1, TAK1 and Tpl2, have been found to stimulate MKK4/7 and JNKs.[145,146] In vitro most of these MAPKKKs stimulate more than one of the MAPK pathways and often the NFκB pathway, too. However, it is rather unclear to which extend this also applies to the in vivo situation. Members of the TNF receptor family activate MAPKKKs and the JNK pathway in concert with adapter proteins of the TNF receptor associated factor (TRAF) group. For example, TNF-R1 and several other receptors recruit TRAF2, which interacts with MEKK1 and ASK1. The JNK signaling pathway has been implicated in proliferation, differentiation and in apoptosis induction, but also in prevention of apoptosis.[145,146]

JNK Signaling and Apoptosis

JNKs and their targets the AP-1 transcription factors fulfill apoptotic and anti-apoptotic functions in vivo, dependent on cell-type and developmental stage. For example, JNK-3 deficient mice display an impaired apoptotic response in models of excitotoxic neuronal cell death, but hepatocytes of embryos deficient in the JNK target cJun undergo massive apoptosis.[147,148] The apoptotic effects of JNK may be mainly related to their capability to upregulate expression of pro-apoptotic proteins, including p53, cMyc, Bim and FasL.[149-156] However, transcription-independent effects of JNKs have also been described. For example, UV-induced apoptosis is mediated by JNKs, but it is only partly inhibited by overexpression of phosphorylation-defect cJun and occurs in the absence of synthesis of new proteins.[157] Noteworthy, the NFκB pathway seems to antagonize the JNK pathway in context of TNF-R1 signaling by upregulation of GADD45β and xIAP.[158,159]

The cJun N-Terminal Kinase Pathway in Fas Signaling

As Fas is mainly recognized as a death inducer the JNK pathway has mainly been analyzed in context of Fas signaling with respect to its proapoptotic properties. In fact, JNK activation was found in several reports when apoptosis was triggered by Fas stimulation. This is in good agreement with other studies showing that caspases can stimulate JNK signaling, e.g., by cleavage and activation of MEKK1.[160-162] An apoptosis-associated caspase-dependent mode of JNK activation fits also well with studies showing that caspase inhibitors prevent both Fas-induced apoptosis and Fas-mediated JNK activation.[160,163-172] Inhibition of JNK signaling does not or only modestly interfere with Fas-mediated apoptosis, indicating that the proapoptotic properties of JNK activation play no essential role in Fas-induced cell death.[163,164,166,167,169,172] JNK signaling might nevertheless enhance and accelerate Fas-mediated apoptosis by mechanisms described above, especially by AP1-mediated upregulation of death ligands including TNF, TRAIL and FasL itself.[155,156] As described in detail in chapter 10, Fas activates caspases without apoptosis induction upon T-cell activation. In fact, DNA binding of AP1 was found to be increased in this situation pointing to the possibility of Fas-induced caspase-mediated JNK signaling under nonapoptotic conditions.[173]

Fas-induced JNK activation can also occur in addition via an alternative FADD/caspase-8-independent pathway which is initiated by interaction of Fas with the receptor-associated protein DAXX and apoptosis signaling kinase-1 (ASK1), a MAP3K capable to induce apoptosis. While DAXX interacts directly with stimulated Fas, ASK1 is recruited to the Fas signaling complex by virtue of its interaction with DAXX.[174,175] As a consequence ASK1 become activated in the DISC. Noteworthy, ASK1-DAXX complexes are located in the cytoplasm and are therefore accessible for Fas, whereas in the absence of ASK1, DAXX is prominently localized in the nucleus, where it interacts with several nuclear proteins including PML, CEN-P and Pax-3 and represses transcription.[176-180] Overexpression of DAXX and ASK1 does not only induce JNK activation but also apoptosis. In contrast to the studies discussed above, inhibition of caspases blocks apoptosis induction, but does not interfere with JNK activation, placing the latter upstream of caspase action.[174] DAXX and FADD interact independently with distinct sites of the Fas death domain.[181] This is evident from a mutant of murine Fas that is unable to interact with FADD due to deletion of a stretch off amino acids located in the N-terminal part of the Fas death domain. This mutant interacts with DAXX and fails to trigger apoptosis, but retains its capability to induce JNK activation. Thus, Fas is able to induce JNK signaling in the absence of apoptosis and caspase activation by a DAXX and ASK1 dependent pathway. The proapoptotic effects associated with the DAXX-ASK1 pathway (e.g., refs. 174,175,182-184) reflect therefore most likely the cytotoxic action of the JNK pathway that can occurs under certain conditions, but does not represent an obligatory step in Fas-induced apoptosis.

Fas-mediated activation of JNK and AP1 in the absence of apoptosis seems to be of special relevance in cardiac hypertrophy, which is observed as a consequence of mechanical stress in

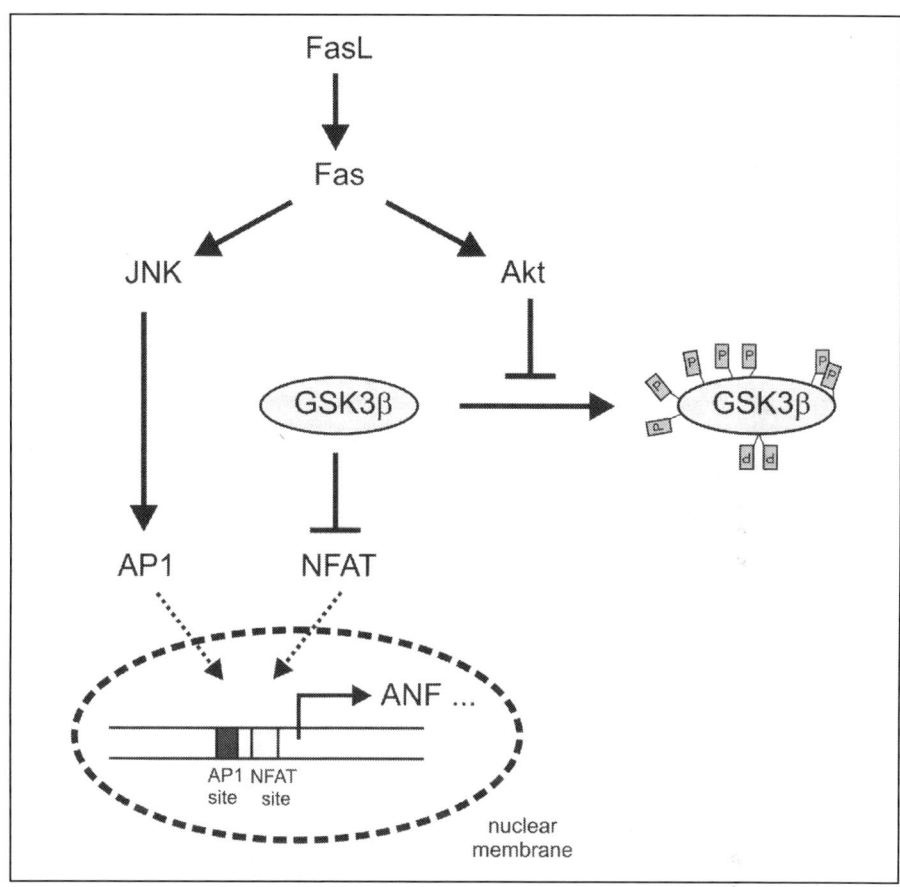

Figure 6. Fas signaling in cardiac hypertrophy. In cultured cardiomyocytes Fas activates by yet unknown mechanisms the Akt and JNK pathways that cooperate in the induction of cardiac hypertrophy e.g., by upregulation of atrial natriuretic factor (ANF) and upregulation of protein synthesis. A central role has glycogen synthase kinase 3β, which inhibits the transcription factor NFAT (nuclear factor of activated T-cells). This kinase becomes inactivated by Akt-mediated phosphorylation after Fas stimulation.

the heart for example in response to arterial hypertension (Fig. 6).[185,186] This growth response is primarily of compensatory and protective nature, but excessive and sustained hypertrophy can also cause heart failure.[187] In recent studies the JNK pathway has been implicated in cardiac hypertrophy in vivo.[188,189] Especially, Fas-mediated JNK activation seems to play a special role in cardiac hypertrophy as this response and JNK activation are diminished in lpr mice.[186] In cultured cardiomyocytes of wild-type mice, but not in cardiomyocytes derived from lpr mice, FasL also activates the PKB/Akt pathway leading to phosphorylation and inhibition of GSK3β, which at least in vitro is necessary for cardiomyocyte hypertrophy.[186,190,191]

FasL and Fas Stimulate Proliferation and Differentiation

Already shortly after the identification of Fas/Apo-1, enhanced proliferation of TCR-stimulated T-cells and thymocytes upon Fas activation has been observed.[192] Especially, studies in T-cell receptor transgenic mice with a nonselecting background and impaired Fas signaling suggested that Fas acts as a costimulator in thymocyte activation during positive selection.[193] Moreover, in human diploid fibroblasts, activation of Fas alone is sufficient to drive proliferation and activation of ERK1/2, a molecular hallmark of proliferation, has been observed in serum-starved fibroblast.[194-197] Fas may also signal apoptosis induction in fibroblast, but the mechanisms / circumstances that determine the quality of the overall Fas response have not been identified so far.[195,196] In particular the molecular architecture of the signaling pathway that leads from Fas stimulation to proliferation in fibroblast is currently completely unknown. In case of Fas-mediated enhancement of TCR-induced T-cell proliferation there is evidence that caspase activation without subsequent apoptosis induction is involved.[173,198-200] Thus, in polyclonal stimulated T-cells TCR-induced proliferation is inhibited by caspase inhibitors.[173,198,199] TCR signaling leads to upregulation of FasL.[201] This opens the possibility that TCR-induced autocrine production of FasL triggers a nonapoptotic caspase-dependent Fas pathway that crucially contributes to T-cell proliferation. This idea would be in agreement with the observation that inhibition of Fas signaling with Fas-Fc inhibits T-cell proliferation induced by suboptimal concentrations of anti-TCR antibodies.[173] Defective T-cell activation as found in FADD-deficient mice and in mice overexpressing dominant-negative FADD would also be in agreement with a role of Fas in TCR-induced proliferation.[202-204] However, T-cell proliferation induced by high concentrations of anti-TCR antibodies was prevented by caspase-inhibitors, while Fas-Fc showed only a marginal effect. This could be related to the comparably poor FasL neutralizing capability of Fas-Fc or could reflect that other death receptors such as TRAIL-R1, TRAIL-R2, DR3 or DR6 can substitute for Fas signaling in TCR-induced T-cell activation. The latter could also explain why Fas-deficient lpr and FasL-deficient gld mice, in contrast to FADD or caspase-8 deficient mice and dominant negative FADD-overexpressing mice, are not impaired in T-cell proliferation.[205,206] However, it is also possible that caspase dependent T-cell proliferation is due to the action of a yet unknown FADD and caspase-8 dependent pathway that is triggered independent of the activation of death receptors. In this case the costimulatory effects of Fas signaling on suboptimally TCR-triggered T-cells could be explained by feeding active caspase-8 in this unknown pathway, which only gains relevance as long the latter is not fully activated. Activation of caspase-8 independent from death receptors can be induced by the caspase-8 interacting proteins BAP31 or Huntingtin-interacting protein HIP-1, but there is no evidence that FADD is involved.[207-209] All published reports consistently found a need of caspase-activity in the first day after TCR stimulation. However, there are contradictory data with respect to the involved caspases and the substrates cleaved. While Kennedy et al found caspase-8 activation but no processing of caspase-3 or the caspase-3 substrate poly(ADP-ribose) polymerase (PARP), Alam et al reported activation of caspases 3, 6, 7 and 8 as well as cleavage of a limited set of caspase-3 substrates.[173,198] They found further cleavage of PARP, but observed, in agreement with a nonapoptotic role of caspase-3 activation, no cleavage of the apoptosis-promoting caspase-3 substrate DNA fragmentation factor (DFF45).[198] The downstream effects of nonapoptotic caspase-activation in T-cell proliferation are largely unknown but seem to involve upregulation of IL2 and caspase-1-mediated IL1 release.[173,210]

It has been recently shown that Fas enhances liver regeneration after partial hepatectomy.[211] In healthy mice Fas activation rapidly induces apoptosis in hepatocytes causing acute liver failure. Thus, during this regenerative response apoptotic Fas signaling has to be blocked e.g., by upregulation of anti-apoptotic proteins. Remarkably, Fas-dependent acceleration of liver regeneration is impaired in lpr mice showing reduced Fas expression, but not in lpr-cg mice, which express Fas that carries a point mutation interfering with the function of death domain.[211] Thus, liver regeneration does not only occur without apoptosis-induction but also requires no intact death domain. The beneficial effect of Fas signaling in liver regeneration correlates with its ability to promote cell cycle progression instead of apoptosis induction in hepatocytes from mice that underwent partial hepatectomy.

Regulation of T-cell activation by the Fas-FasL system gains additional complexity with the observation that the membrane FasL molecule itself retrogradely signals either proliferation or cell cycle arrest in T-cells. When CD8+ T-cells were suboptimally triggered with anti-CD3, proliferation was inhibited by Fas-Fc, a fusion protein containing the extracellular domain of Fas and the Fc portion of IgG (Fas-IgG). More importantly, upon immobilization on plastic, the same reagent costimulated proliferation of wt and Fas deficient CD8+ T-cells, but showed no effect on FasL-deficient cells suggesting that FasL actively signals in a retrograde manner.[212,213] In CD4+ T-cells retrograde FasL signaling has also been demonstrated leading to such diverse effects as cell cycle arrest [68] and proliferation.[214,215] The mechanisms that finally regulate the outcome of retrograde FasL signaling in CD4+ T-cells, are however, still unclear. For a more detailed description of retrograde signaling and a discussion of the intracellular signaling pathways involved please see the chapter by Linkermann et al.

Fas and FasL in Inflammation

FasL expression and FasL-induced T-cell killing have been discussed as a mechanism contributing to the suppression of inflammation-related destruction of certain tissues, so called immune privileged sites, e.g., the cornea of the eye, testis, placenta or pregnant uterus.[216] With the idea of immune privilege in mind several groups have tried to prevent allograft rejection by the immune system of the recipient by ectopic expression of FasL in the transplant. In a limited number of studies this strategy seems to work. For example the inhibition of islet allograft rejection by cotransplantation with FasL expressing myoblasts has been reported.[217] However, in most cases FasL failed to mediate protection to grafted islets or tumor cells and even enhanced rejection.[218-223] Although, transplants elicit a T- and B-cell response these cells appear to be dispensable for FasL-dependent rejection. The development of granulocytic/neutrophilic infiltrates has also been described in FasL-mediated rejection of transplants, but the functional relevance of this observation is not clear yet. FasL-induced recruitment of neutrophils seems to be complex and may be based on several mechanisms that may act redundantly. While some reports argued for a direct chemotactic function of shedded soluble FasL, other studies failed to confirm this finding and instead suggested that neutrophil attractive chemokines, e.g., IL8 and MCP-1, are induced by Fas.[224-229] Indeed, it has been found in various cell types that Fas activation induces the production of IL8 and MCP-1, but the signaling pathway(s) involved in this response are poorly understood. Fas-induced chemokine production has shown to be the consequence of apoptosis induction and caspase activation in some cases, but can also take place in the absence of apoptosis and caspase activation (Fig. 7). In the latter case this might be due to Fas-induced activation of the NFκB pathway by mechanisms described under 6.2.5. In addition, Fas may indirectly trigger chemokine production independent from its own capability to directly induce proinflammatory signaling pathways, as it can enhance LPS- and IL1-induced NFκB activation in macrophages and dendritic cells. This depends on the capability of FADD to interact with MyD88, an essential component of IL1R/TLR signaling. Overexpression of FADD inhibits LPS- and IL1-induced NFκB activation.[120,121] The FADD-MyD88 complex seems to be therefore inappropriate for IL1R/TLR signaling. FADD recruitment to activated Fas complexes is accompanied by MyD88 release from FADD-MyD88 complexes. The resulting increase of "free" MyD88 that is available for IL1R and TLRs may then boost IL1R/TLR signaling.[120] Thus, in context of direct Fas-induced NFκB activation FADD fulfils a NFκB-inducing function as described under 6.2.5. However, in context of IL1R/TLR signaling FADD has a NFκB inhibiting role. In FLIP-L expressing cells where Fas-induced activation of NFκB and apoptosis induction are blocked, Fas may therefore exert a proinflammatory effect by antagonizing the inhibitory effect of FADD in IL1R/TLR signaling.

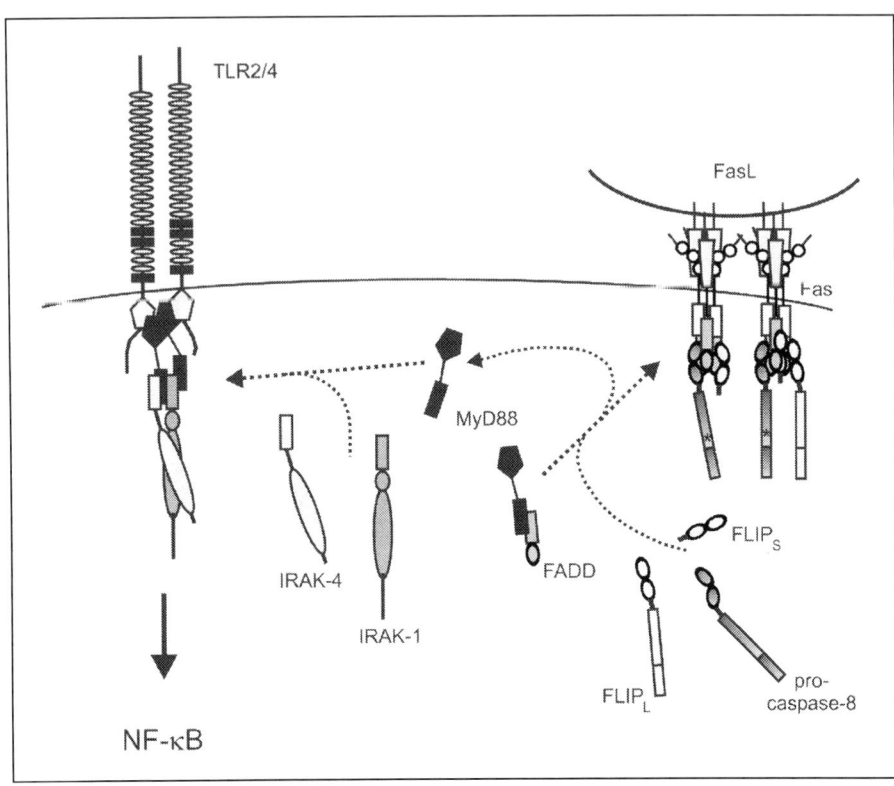

Figure 7. Mechanisms of Fas-mediated inflammation. FADD forms complexes with MyD88, a crucial signaling intermediate of the Interleukin-1 receptor/Toll-like receptor (IL1R/TLR) signaling pathway. The FADD-MyD88 complexes are not able to support IL1R/TLR signaling. Fas stimulation enhances LPS (or IL1) induced NFκB signaling by recruitment of FADD which is accompanied by resolution of FADD-MyD88 complexes. In cells where Fas-induced apoptosis and NFκB activation are blocked by FLIP proteins, this leads to Fas-mediated enhancement of LPS (IL1) induced NFκB activation by FasL without apparent NFκB activation by stimulation of Fas alone.

Fas-Induced ERK Activation

Fas and FasL are upregulated in the brain in a variety of disorders and have been implicated in apoptosis induction in neurons of the central nervous system (CNS) after cerebral hypoxia/ischeamic insult and in developing motoneurons after target contact.[230] Despite promiscuous expression of Fas and FasL in the brain, Fas induces no cell death in most cell types of the brain including dorsal root ganglion (DRG) cortical neurons, cerebral cortical neurons and astrocytes. Thus, Fas may fulfil additional nonapoptotic functions in the brain. In this regard it has been shown that Fas signals ERK activation in neuronal cell lines and induces neurite outgrowth in DRG explants.[231,232] Moreover, a beneficial role of endogenous Fas signaling has been deduced from studies in lpr mice showing reduced regeneration after sciatic nerve crush injury. In one study, Fas-induced activation of the MEK1-ERK pathway was dependent on caspase activation whereas another report showed that ERK activation and neurite outgrowth in DRGs occur independent from caspases. In agreement with the latter, neurite outgrowth was normal in this study in DRGs derived from lpr-cg mice, expressing a Fas mutant with a defective death domain.[232] In this regard Fas-induced neurite outgrowth seems to be mechanistically related to Fas-mediated acceleration of liver regeneration. A further report showed FLIP dependent recruitment of Raf-1, an ERK pathway activating MAP3K, to Fas.[233] This would be in accordance with a caspase-independent mode of ERK activation, but is contradictory in some aspects as FLIP-L mediated recruitment of Raf-1 to Fas should require in this case an intact death domain in Fas to allow FADD-mediated recruitment of and FLIP-L and Raf-1. Fas-induced ERK activation has also been implicated in Fas-induced maturation of dendritic cells and in Fas-induced proliferation of serum starved fibroblasts.[197,234] Future studies must reveal to which extend Fas utilize different means to activate the MEK1-ERK pathway and /or whether the described aspects of Fas-induced ERK activation reflect properties of one or more redundant pathways.

Acknowledgement

This work was supported by Deutsche Forschungsgemeinschaft (Grant Wa 1025/11-1 Sonderforschungsbereich 495 project A5) and Dr. Mildred Scheel Stiftung für Krebsforschung grant 10-1751.

References

1. Tanaka M, Suda T, Takahashi T et al. Expression of the functional soluble form of human fas ligand in activated lymphocytes. EMBO J 1995; 14:1129-1135.
2. Kayagaki N, Kawasaki A, Ebata T et al. Metalloproteinase-mediated release of human Fas ligand. J Exp Med 1995; 182:1777-1783.
3. Suda T, Hashimoto H, Tanaka M et al. Membrane fas ligand kills human peripheral blood T lymphocytes, and soluble Fas ligand blocks the killing. J Exp Med 1997; 186:2045-2050.
4. Tanaka M, Itai T, Adachi M et al. Regulation of Fas ligand by shedding. Nat Med 1998; 4:31-36.
5. Schneider P, Holler N, Bodmer JL et al. Conversion of membrane-bound Fas(CD95) ligand to its soluble form is associated with downregulation of its proapoptotic activity and loss of liver toxicity. J Exp Med 1998; 187:1205-1213.
6. Itoh N, Yonehara S, Ishii A et al. The polypeptide encoded by the cDNA for human cell surface antigen Fas can mediate apoptosis. Cell 1991; 66:233-243.
7. Itoh N, Nagata S. A novel protein domain required for apoptosis. Mutational analysis of human Fas antigen. J Biol Chem 1993; 268:10932-10937.
8. Fesik SW. Insights into programmed cell death through structural biology. Cell 2000; 103:273-282.
9. Papoff G, Hausler P, Eramo A et al. Identification and characterization of a ligand-independent oligomerization domain in the extracellular region of the CD95 death receptor. J Biol Chem 1999; 274:38241-38250.
10. Chan FK, Chun HJ, Zheng L et al. A domain in TNF receptors that mediates ligand-independent receptor assembly and signaling. Science 2000; 288:2351-2354.
11. Siegel RM, Frederiksen JK, Zacharias DA et al. Fas preassociation required for apoptosis signaling and dominant inhibition by pathogenic mutations. Science 2000; 288:2354-2357.
12. Kischkel FC, Hellbardt S, Behrmann I et al. Cytotoxicity-dependent APO-1 (Fas/CD95)-associated proteins form a death-inducing signaling complex (DISC) with the receptor. EMBO J 1995; 14:5579-5588.
13. Muzio M, Chinnaiyan AM, Kischkel FC et al. FLICE, a novel FADD-homologous ICE/CED-3-like protease, is recruited to the CD95 (Fas/APO-1) death—inducing signaling complex. Cell 1996; 85:817-827.
14. Chinnaiyan AM, O'Rourke K, Tewari M et al. A novel death domain-containing protein, interacts with the death domain of Fas and initiates apoptosis. Cell 1995; 81:505-512.
15. Boldin MP, Varfolomeev EE, Pancer Z et al. A novel protein that interacts with the death domain of Fas/APO1 contains a sequence motif related to the death domain. J Biol Chem 1995; 270:7795-7798.
16. Donepudi M, Mac Sweeney A, Briand C et al. Insights into the regulatory mechanism for caspase-8 activation. Mol Cell 2003; 11:543-549.

17. Boatright KM, Renatus M, Scott FL et al. A unified model for apical caspase activation. Mol Cell 2003; 11:529-541.
18. Peter ME, Krammer PH. The CD95(APO-1/Fas) DISC and beyond. Cell Death Differ 2003; 10:26-35.
19. Barnhart BC, Alappat EC, Peter ME. The CD95 type I/type II model. Semin Immunol 2003; 15:185-193.
20. Vaux DL, Silke J. Mammalian mitochondrial IAP binding proteins. Biochem Biophys Res Commun 2003; 304:499-504.
21. Shi Y. Apoptosome: The cellular engine for the activation of caspase-9. Structure (Camb) 2002; 10:285-288.
22. Holler N, Zaru R, Micheau O et al. Fas triggers an alternative, caspase-8-independent cell death pathway using the kinase RIP as effector molecule. Nat Immunol 2000; 1:489-495.
23. Matsumura H, Shimizu Y, Ohsawa Y et al. Necrotic death pathway in Fas receptor signaling. J Cell Biol 2000; 151:1247-1256.
24. Vercammen D, Brouckaert G, Denecker G et al. Dual signaling of the Fas receptor: Initiation of both apoptotic and necrotic cell death pathways. J Exp Med 1998; 188:919-930.
25. Silverman N, Maniatis T. NF-kappaB signaling pathways in mammalian and insect innate immunity. Genes Dev 2001; 15:2321-2342.
26. Li Q, Verma IM. NF-kappaB regulation in the immune system. Nat Rev Immunol 2002; 2:725-734.
27. Huxford T, Huang DB, Malek S et al. The crystal structure of the IkappaBalpha/NF-kappaB complex reveals mechanisms of NF-kappaB inactivation. Cell 1998; 95:759-770.
28. Jacobs MD, Harrison SC. Structure of an IkappaBalpha/NF-kappaB complex. Cell 1998; 95:749-758.
29. Brown K, Gerstberger S, Carlson L et al. Control of I kappa B-alpha proteolysis by site-specific, signal-induced phosphorylation. Science 1995; 267:1485-1488.
30. Brockman JA, Scherer DC, McKinsey TA et al. Coupling of a signal response domain in I kappa B alpha to multiple pathways for NF-kappa B activation. Mol Cell Biol 1995; 15:2809-2818.
31. DiDonato J, Mercurio F, Rosette C et al. Mapping of the inducible IkappaB phosphorylation sites that signal its ubiquitination and degradation. Mol Cell Biol 1996; 16:1295-1304.
32. Yaron A, Hatzubai A, Davis M et al. Identification of the receptor component of the IkappaBalpha-ubiquitin ligase. Nature 1998; 396:590-594.
33. Baldi L, Brown K, Franzoso G et al. Critical role for lysines 21 and 22 in signal-induced, ubiquitin-mediated proteolysis of I kappa B-alpha. J Biol Chem 1996; 271:376-379.
34. Scherer DC, Brockman JA, Chen Z et al. Signal-induced degradation of I kappa B alpha requires site-specific ubiquitination. Proc Natl Acad Sci USA 1995; 92:11259-11263.
35. Rice NR, MacKichan ML, Israel A. The precursor of NF-kappa B p50 has I kappa B-like functions. Cell 1992; 71:243-253.
36. Naumann M, Wulczyn FG, Scheidereit C. The NF-kappa B precursor p105 and the proto-oncogene product Bcl-3 are I kappa B molecules and control nuclear translocation of NF-kappa B. EMBO J 1993; 12:213-222.
37. Mercurio F, DiDonato JA, Rosette C et al. p105 and p98 precursor proteins play an active role in NF-kappa B-mediated signal transduction. Genes Dev 1993; 7:705-718.
38. Dobrzanski P, Ryseck RP, Bravo R. Specific inhibition of RelB/p52 transcriptional activity by the C-terminal domain of p100. Oncogene 1995; 10:1003-1007.
39. Heusch M, Lin L, Geleziunas R et al. The generation of nfkb2 p52: Mechanism and efficiency. Oncogene 1999; 18:6201-6208.
40. Lin L, Ghosh S. A glycine-rich region in NF-kappaB p105 functions as a processing signal for the generation of the p50 subunit. Mol Cell Biol 1996; 16:2248-2254.
41. Orian A, Schwartz AL, Israel A et al. Structural motifs involved in ubiquitin-mediated processing of the NF-kappaB precursor p105: Roles of the glycine-rich region and a downstream ubiquitination domain. Mol Cell Biol 1999; 19:3664-3673.
42. Rothwarf DM, Karin M. The NF-kappa B activation pathway: A paradigm in information transfer from membrane to nucleus. Sci STKE 1999; 1999:RE1.
43. Sigala JL, Bottero V, Young DB et al. Activation of transcription factor NF-kappaB requires ELKS, an IkappaB kinase regulatory subunit. Science 2004; 304:1963-1967.

44. Chen G, Cao P, Goeddel DV. TNF-induced recruitment and activation of the IKK complex require Cdc37 and Hsp90. Mol Cell 2002; 9:401-410.
45. Wajant H, Henkler F, Scheurich P. The TNF-receptor-associated factor family: Scaffold molecules for cytokine receptors, kinases and their regulators. Cell Signal 2001; 13:389-400.
46. Lee FS, Peters RT, Dang LC et al. MEKK1 activates both IkappaB kinase alpha and IkappaB kinase beta. Proc Natl Acad Sci USA 1998; 95:9319-9324.
47. Nakano H, Shindo M, Sakon S et al. Differential regulation of IkappaB kinase alpha and beta by two upstream kinases, NF-kappaB-inducing kinase and mitogen-activated protein kinase/ERK kinase kinase-1. Proc Natl Acad Sci USA 1998; 95:3537-3542.
48. Zhao Q, Lee FS. Mitogen-activated protein kinase/ERK kinase kinases 2 and 3 activate nuclear factor-kappaB through IkappaB kinase-alpha and IkappaB kinase-beta. J Biol Chem 1999; 274:8355-8358.
49. Ninomiya-Tsuji J, Kishimoto K, Hiyama A et al. The kinase TAK1 can activate the NIK-I kappaB as well as the MAP kinase cascade in the IL-1 signalling pathway. Nature 1999; 398:252-256.
50. Sizemore N, Leung S, Stark GR. Activation of phosphatidylinositol 3-kinase in response to interleukin-1 leads to phosphorylation and activation of the NF-kappaB p65/RelA subunit. Mol Cell Biol 1999; 19:4798-4805.
51. Madrid LV, Wang CY, Guttridge DC et al. Akt suppresses apoptosis by stimulating the transactivation potential of the RelA/p65 subunit of NF-kappaB. Mol Cell Biol 2000; 20:1626-1638.
52. Leitges M, Sanz L, Martin P et al. Targeted disruption of the zetaPKC gene results in the impairment of the NF-kappaB pathway. Mol Cell 2001; 8:771-7808.
53. Duran A, Diaz-Meco MT, Moscat J. Essential role of RelA Ser311 phosphorylation by zetaPKC in NF-kappaB transcriptional activation. EMBO J 2003; 22:3910-3918.
54. Zhong H, SuYang H, Erdjument-Bromage H et al. Transcriptional activity of NF-kappaB is regulated by the IkappaB-associated PKAc subunit through a cyclic AMP-independent mechanism. Cell 1997; 89:413-424.
55. Wang D, Westerheide SD, Hanson JL et al. Tumor necrosis factor alpha-induced phosphorylation of RelA/p65 on Ser529 is controlled by casein kinase II. J Biol Chem 2000; 275:32592-32597.
56. Sakurai H, Chiba H, Miyoshi H et al. IkappaB kinases phosphorylate NF-kappaB p65 subunit on serine 536 in the transactivation domain. J Biol Chem 1999; 274:30353-30356.
57. Rudolph D, Yeh WC, Wakeham A et al. Severe liver degeneration and lack of NF-kappaB activation in NEMO/IKKgamma-deficient mice. Genes Dev 2000; 14:854-862.
58. Schmidt-Supprian M, Bloch W, Courtois G et al. NEMO/IKK gamma-deficient mice model incontinentia pigmenti. Mol Cell 2000; 5:981-992.
59. Makris C, Godfrey VL, Krahn-Senftleben G et al. Female mice heterozygous for IKK gamma/NEMO deficiencies develop a dermatopathy similar to the human X-linked disorder incontinentia pigmenti. Mol Cell 2000; 5:969-979.
60. Salmeron A, Janzen J, Soneji Y et al. Direct phosphorylation of NF-kappaB1 p105 by the IkappaB kinase complex on serine 927 is essential for signal-induced p105 proteolysis. J Biol Chem 2001; 276:22215-22222.
61. Yamaoka S, Courtois G, Bessia C et al. Complementation cloning of NEMO, a component of the IkappaB kinase complex essential for NF-kappaB activation. Cell 1998; 93:1231-1240.
62. Harhaj EW, Good L, Xiao G et al. Somatic mutagenesis studies of NF-kappa B signaling in human T cells: Evidence for an essential role of IKK gamma in NF-kappa B activation by T-cell costimulatory signals and HTLV-I Tax protein. Oncogene 2000; 19:1448-1456.
63. Tanaka M, Fuentes ME, Yamaguchi K et al. Embryonic lethality, liver degeneration, and impaired NF-kappa B activation in IKK-beta-deficient mice. Immunity 1999; 10:421-429.
64. Li Q, Lu Q, Hwang JY et al. IKK1-deficient mice exhibit abnormal development of skin and skeleton. Genes Dev 1999; 13:1322-1328.
65. Takeda K, Takeuchi O, Tsujimura T et al. Limb and skin abnormalities in mice lacking IKKalpha. Science 1999; 284:313-316.
66. Hu Y, Baud V, Delhase M et al. Abnormal morphogenesis but intact IKK activation in mice lacking the IKKalpha subunit of IkappaB kinase. Science 1999; 284:316-320.
67. Li Q, Van Antwerp D, Mercurio F et al. Severe liver degeneration in mice lacking the IkappaB kinase 2 gene. Science 1999; 284:321-325.

68. Yamamoto Y, Verma UN, Prajapati S et al. Histone H3 phosphorylation by IKK-alpha is critical for cytokine-induced gene expression. Nature 2003; 423:655-659.

69. Anest V, Hanson JL, Cogswell PC et al. A nucleosomal function for IkappaB kinase-alpha in NF-kappaB-dependent gene expression. Nature 2003; 423:659-663.

70. Dejardin E, Droin NM, Delhase M et al. The lymphotoxin-beta receptor induces different patterns of gene expression via two NF-kappaB pathways. Immunity 2002; 17:525-535.

71. Claudio E, Brown K, Park S et al. BAFF-induced NEMO-independent processing of NF-kappa B2 in maturing B cells. Nat Immunol 2002; 3:958-965.

72. Kayagaki N, Yan M, Seshasayee D et al. BAFF/BLyS receptor 3 binds the B cell survival factor BAFF ligand through a discrete surface loop and promotes processing of NF-kappaB2. Immunity 2002; 17:515-524.

73. Coope HJ, Atkinson PG, Huhse B et al. CD40 regulates the processing of NF-kappaB2 p100 to p52. EMBO J 2002; 21:5375-5385.

74. Saitoh T, Nakayama M, Nakano H et al. TWEAK induces NF-kappaB2 p100 processing and long lasting NF-kappaB activation. J Biol Chem 2003; 278:36005-36012.

75. Xiao G, Cvijic ME, Fong A et al. Retroviral oncoprotein tax induces processing of NF-kappaB2/ p100 in T cells: Evidence for the involvement of IKKalpha. EMBO J 2001; 20:6805-6815.

76. Senftleben U, Cao Y, Xiao G et al. Activation by IKKalpha of a second, evolutionary conserved, NF-kappa B signaling pathway. Science 2001; 293:1495-1499.

77. Xiao G, Harhaj EW, Sun SC. NF-kappaB-inducing kinase regulates the processing of NF-kappaB2 p100. Mol Cell 2001; 7:401-409.

78. Fong A, Sun SC. Genetic evidence for the essential role of beta-transducin repeat-containing protein in the inducible processing of NF-kappa B2/p100. J Biol Chem 2002; 277:22111-22114.

79. Delhase M, Hayakawa M, Chen Y et al. Positive and negative regulation of IkappaB kinase activity through IKKbeta subunit phosphorylation. Science 1999; 284:309-313.

80. Arenzana-Seisdedos F, Thompson J, Rodriguez MS et al. Inducible nuclear expression of newly synthesized I kappa B alpha negatively regulates DNA-binding and transcriptional activities of NF-kappa B. Mol Cell Biol 1995; 15:2689-2696.

81. Arenzana-Seisdedos F, Turpin P, Rodriguez M et al. Nuclear localization of I kappa B alpha promotes active transport of NF-kappa B from the nucleus to the cytoplasm. J Cell Sci 1997; 110:369-378.

82. Sachdev S, Hoffmann A, Hannink M. Nuclear localization of IkappaB alpha is mediated by the second ankyrin repeat: The IkappaB alpha ankyrin repeats define a novel class of cis-acting nuclear import sequences. Mol Cell Biol 1998; 18:2524-2534.

83. Sachdev S, Bagchi S, Zhang DD et al. Nuclear import of IkappaBalpha is accomplished by a ran-independent transport pathway. Mol Cell Biol 2000; 20:1571-1582.

84. Suyang H, Phillips R, Douglas I et al. Role of unphosphorylated, newly synthesized I kappa B beta in persistent activation of NF-kappa B. Mol Cell Biol 1996; 16:5444-5449.

85. DeLuca C, Petropoulos L, Zmeureanu D et al. Nuclear IkappaBbeta maintains persistent NF-kappaB activation in HIV-1-infected myeloid cells. J Biol Chem 1999; 274:13010-13016.

86. He KL, Ting AT. A20 inhibits tumor necrosis factor (TNF) alpha-induced apoptosis by disrupting recruitment of TRADD and RIP to the TNF receptor 1 complex in Jurkat T cells. Mol Cell Biol 2002; 22:6034-6045.

87. Wertz IE, O'Rourke KM, Zhou H et al. De-ubiquitination and ubiquitin ligase domains of A20 downregulate NF-kappaB signalling. Nature 2004.

88. Kobayashi K, Hernandez LD, Galan JE et al. IRAK-M is a negative regulator of Toll-like receptor signaling. Cell 2002; 110:191-202.

89. Janssens S, Burns K, Tschopp J et al. Regulation of interleukin-1- and lipopolysaccharide-induced NF-kappaB activation by alternative splicing of MyD88. Curr Biol 2002; 12:467-471.

90. Beg AA, Sha WC, Bronson RT et al. Embryonic lethality and liver degeneration in mice lacking the RelA component of NF-kappa B. Nature 1995; 376:167-170.

91. Beg AA, Baltimore D. An essential role for NF-kappaB in preventing TNF-alpha-induced cell death. Science 1996; 274:782-784.

92. Van Antwerp DJ, Martin SJ, Kafri T et al. Suppression of TNF-alpha-induced apoptosis by NF-kappaB. Science 1996; 274:787-789.

93. Baldwin AS. Control of oncogenesis and cancer therapy resistance by the transcription factor NF-kappaB. J Clin Invest 2001; 107:241-246.

94. Wang CY, Mayo MW, Korneluk RG et al. NF-kappaB antiapoptosis: Induction of TRAF1 and TRAF2 and c-IAP1 and c-IAP2 to suppress caspase-8 activation. Science 1998; 281:1680-1683.

95. Chu ZL, McKinsey TA, Liu L et al. Suppression of tumor necrosis factor-induced cell death by inhibitor of apoptosis c-IAP2 is under NF-kappaB control. Proc Natl Acad Sci USA 1997; 94:10057-10062.

96. Stehlik C, de Martin R, Kumabashiri I et al. Nuclear factor (NF)- kappaB-regulated X-chromosome-linked iap gene expression protects endothelial cells from tumor necrosis factor alpha-induced apoptosis. J Exp Med 1998; 188:211-216.

97. Schwenzer R, Siemienski K, Liptay S et al. The human tumor necrosis factor (TNF) receptor-associated factor 1 gene (TRAF1) is up-regulated by cytokines of the TNF ligand family and modulates TNF-induced activation of NF-kappaB and c-Jun N-terminal kinase. J Biol Chem 1999; 274:19368-19374.

98. Kreuz S, Siegmund D, Scheurich P et al. NF- kappaB inducers upregulate cFLIP, a cycloheximide-sensitive inhibitor of death receptor signaling. Mol Cell Biol 2001; 21:3964-3973.

99. Micheau O, Lens S, Gaide O et al. NF-kappaB signals induce the expression of c-FLIP. Mol Cell Biol 2001; 21:5299-5305.

100. You Z, Ouyang H, Lopatin D et al. Nuclear factor-kappa B-inducible death effector domain-containing protein suppresses tumor necrosis factor-mediated apoptosis by inhibiting caspase-8 activity. J Biol Chem 2001; 276:26398-26404.

101. Zong WX, Edelstein LC, Chen C et al. The prosurvival Bcl-2 homolog Bfl-1/A1 is a direct transcriptional target of NF-kappaB that blocks TNFalpha-induced apoptosis. Genes Dev 1999; 13:382-387.

102. Wang CY, Guttridge DC, Mayo MW et al. NF-kappaB induces expression of the Bcl-2 homologue A1/Bfl-1 to preferentially suppress chemotherapy-induced apoptosis. Mol Cell Biol 1999; 19:5923-5929.

103. Lee HH, Dadgostar H, Cheng Q et al. NF-kappaB-mediated up-regulation of Bcl-x and Bfl-1/A1 is required for CD40 survival signaling in B lymphocytes. Proc Natl Acad Sci USA 1999; 96:9136-9141.

104. Chen C, Edelstein LC, Gelinas C. The Rel/NF-kappaB family directly activates expression of the apoptosis inhibitor Bcl-x(L). Mol Cell Biol 2000; 20:2687-2695.

105. Bernard D, Quatannens B, Vandenbunder B et al. Rel/NF-kappaB transcription factors protect against tumor necrosis factor (TNF)-related apoptosis-inducing ligand (TRAIL)-induced apoptosis by up-regulating the TRAIL decoy receptor DcR1. J Biol Chem 2001; 276:27322-27328.

106. Lin Y, Devin A, Rodriguez Y et al. Cleavage of the death domain kinase RIP by caspase-8 prompts TNF-induced apoptosis. Genes Dev 1999; 13:2514-2526.

107. Martinon F, Holler N, Richard C et al. Activation of a pro-apoptotic amplification loop through inhibition of NF-kappaB-dependent survival signals by caspase-mediated inactivation of RIP. FEBS Lett 2000; 468:134-136.

108. Irmler M, Steiner V, Ruegg C et al. Caspase-induced inactivation of the anti-apoptotic TRAF1 during Fas ligand-mediated apoptosis. FEBS Lett 2000; 468:129-133.

109. Barkett M, Xue D, Horvitz HR et al. Phosphorylation of IkappaB-alpha inhibits its cleavage by caspase CPP32 in vitro. J Biol Chem 1997; 272:29419-29422.

110. Tang G, Yang J, Minemoto Y et al. Blocking caspase-3-mediated proteolysis of IKKbeta suppresses TNF-alpha-induced apoptosis. Mol Cell 2001; 8:1005-1016.

111. Bachelder RE, Ribick MJ, Marchetti A et al. p53 inhibits alpha 6 beta 4 integrin survival signaling by promoting the caspase 3-dependent cleavage of AKT/PKB. J Cell Biol 1999; 147:1063-1072.

112. Hu WH, Johnson H, Shu HB. Activation of NF-kappaB by FADD, casper, and caspase-8. J Biol Chem 2000; 275:10838-10844.

113. Arnold R, Liou J, Drexler HC et al. Caspase-mediated cleavage of hematopoietic progenitor kinase 1 (HPK1) converts an activator of NFkappaB into an inhibitor of NFkappaB. J Biol Chem 2001; 276:14675-14684.

114. Ravi R, Bedi A, Fuchs EJ. CD95 (Fas)-induced caspase-mediated proteolysis of NF-kappaB. Cancer Res 1998; 58:882-886.

115. Levkau B, Scatena M, Giachelli CM et al. Apoptosis overrides survival signals through a caspase-mediated dominant-negative NF-kappa B loop. Nat Cell Biol 1999; 1:227-233.
116. Wajant H, Haas E, Schwenzer R et al. Inhibition of death receptor-mediated gene induction by a cycloheximide-sensitive factor occurs at the level of or upstream of Fas-associated death domain protein (FADD). J Biol Chem 2000; 275:24357-2436611.
117. Krippner-Heidenreich A, Tubing F, Bryde S et al. Control of receptor-induced signaling complex formation by the kinetics of ligand/receptor interaction. J Biol Chem 2002; 277:44155-44163.
118. Chaudhary PM, Eby MT, Jasmin A et al. Activation of the NF-kappaB pathway by caspase 8 and its homologs. Oncogene 2000; 19:4451-4460.
119. Kreuz S, Siegmund D, Rumpf JJ et al. NFkappa B activation by Fas is mediated through FADD, Caspase-8 and RIP and is inhibited by FLIP. J Cell Biology 2004; 166:369-380. in press.
120. Ma Y, Liu H, Tu-Rapp H et al. Fas ligation on macrophages enhances IL-1R1-Toll-like receptor 4 signaling and promotes chronic inflammation. Nat Immunol 2004; 5:380-387.
121. Bannerman DD, Tupper JC, Kelly JD et al. The Fas-associated death domain protein suppresses activation of NF-kappa B by LPS and IL-1 beta. J Clin Invest 2002; 109:419-425.
122. Stanger BZ, Leder P, Lee TH et al. RIP: A novel protein containing a death domain that interacts with Fas/APO-1 (CD95) in yeast and causes cell death. Cell 1995; 81:513-523.
123. Ting AT, Pimentel-Muinos FX, Seed B. RIP mediates tumor necrosis factor receptor 1 activation of NF-kappaB but not Fas/APO-1-initiated apoptosis. EMBO J 1996; 15:6189-6196.
124. Kelliher MA, Grimm S, Ishida Y et al. The death domain kinase RIP mediates the TNF-induced NF-kappaB signal. Immunity 1998; 8:297-303.
125. Lin Y, Devin A, Cook A et al. The death domain kinase RIP is essential for TRAIL (Apo2L)-induced activation of IkappaB kinase and c-Jun N-terminal kinase. Mol Cell Biol 2000; 20:6638-6645.
126. Leverkus M, Neumann M, Mengling T et al. Regulation of tumor necrosis factor-related apoptosis-inducing ligand sensitivity in primary and transformed human keratinocytes. Cancer Res 2000; 60:553-559.
127. Fulda S, Meyer E, Debatin KM. Metabolic inhibitors sensitize for CD95 (APO-1/Fas)-induced apoptosis by down-regulating Fas-associated death domain-like interleukin 1-converting enzyme inhibitory protein expression. Cancer Res 2000; 60:3947-3956.
128. Chang DW, Xing Z, Pan Y et al. c-FLIP(L) is a dual function regulator for caspase-8 activation and CD95-mediated apoptosis. EMBO J 2002; 21:3704-3714.
129. Wajant H, Pfizenmaier K, Scheurich P. Tumor necrosis factor signaling. Cell Death Differ 2003; 10:45-65.
130. Li Z, Zhang J, Chen D et al. Casper/c-FLIP is physically and functionally associated with NF-kappaB1 p105. Biochem Biophys Res Commun 2003; 309:980-985.
131. Grambihler A, Higuchi H, Bronk SF et al. cFLIP-L inhibits p38 MAPK activation: An additional anti-apoptotic mechanism in bile acid-mediated apoptosis. J Biol Chem 2003; 278:26831-26837.
132. Fang LW, Tai TS, Yu WN et al. Phosphatidylinositide 3-kinase priming couples c-FLIP to T cell activation. J Biol Chem 2004; 279:13-18.
133. Lens SM, Kataoka T, Fortner KA et al. The caspase 8 inhibitor c-FLIP(L) modulates T-cell receptor-induced proliferation but not activation-induced cell death of lymphocytes. Mol Cell Biol 2002; 22:5419-5433.
134. Tai TS, Fang LW, Lai MZ et al. c-FLICE inhibitory protein expression inhibits T-cell activation. Cell Death Differ 2004; 11:69-79.
135. Wu W, Rinaldi L, Fortner KA et al. Cellular FLIP long form-transgenic mice manifest a Th2 cytokine bias and enhanced allergic airway inflammation. J Immunol 2004; 172:4724-4732.
136. Liu L, Eby MT, Rathore N et al. The human herpes virus 8-encoded viral FLICE inhibitory protein physically associates with and persistently activates the Ikappa B kinase complex. J Biol Chem 2002; 277:13745-13751.
137. Sun Q, Matta H, Chaudhary PM. The human herpes virus 8-encoded viral FLICE inhibitory protein protects against growth factor withdrawal-induced apoptosis via NF-kappa B activation. Blood 2003; 101:1956-1961.
138. Field N, Low W, Daniels M et al. KSHV vFLIP binds to IKK-gamma to activate IKK. J Cell Sci 2003; 116:3721-3728.

139. Sun Q, Zachariah S, Chaudhary PM. The human herpes virus 8-encoded viral FLICE-inhibitory protein induces cellular transformation via NF-kappaB activation. J Biol Chem 2003; 278:52437-52445.
140. Matta H, Sun Q, Moses G et al. Molecular genetic analysis of human herpes virus 8-encoded viral FLICE inhibitory protein-induced NF-kappaB activation. J Biol Chem 2003; 278:52406-52411.
141. Hopkins-Donaldson S, Bodmer JL, Bourloud KB et al. Loss of caspase-8 expression in highly malignant human neuroblastoma cells correlates with resistance to tumor necrosis factor-related apoptosis-inducing ligand-induced apoptosis. Cancer Res 2000; 60:4315-4319.
142. Teitz T, Wei T, Valentine MB et al. Caspase 8 is deleted or silenced preferentially in childhood neuroblastomas with amplification of MYCN. Nat Med 2000; 6:529-535.
143. Takita J, Yang HW, Bessho F et al. Absent or reduced expression of the caspase 8 gene occurs frequently in neuroblastoma, but not commonly in Ewing sarcoma or rhabdomyosarcoma. Med Pediatr Oncol 2000; 35:541-543.
144. Eggert A, Grotzer MA, Zuzak TJ et al. Resistance to TRAIL-induced apoptosis in neuroblastoma cells correlates with a loss of caspase-8 expression. Med Pediatr Oncol 2000; 35:603-607.
145. Davis RJ. Signal transduction by the JNK group of MAP kinases. Cell 2000; 103:239-252.
146. Shaulian E, Karin M. AP-1 as a regulator of cell life and death. Nat Cell Biol 2002; 4:131-136.
147. Yang DD, Kuan CY, Whitmarsh AJ et al. Absence of excitotoxicity-induced apoptosis in the hippocampus of mice lacking the Jnk3 gene. Nature 1997; 389:865-870.
148. Hilberg F, Aguzzi A, Howells N et al. c-jun is essential for normal mouse development and hepatogenesis. Nature 1993; 365:179-181.
149. Fuchs SY, Adler V, Pincus MR et al. MEKK1/JNK signaling stabilizes and activates p53. Proc Natl Acad Sci USA 1998; 95:10541-10546.
150. Fuchs SY, Adler V, Buschmann T et al. JNK targets p53 ubiquitination and degradation in nonstressed cells. Genes Dev 1998; 12:2658-2663.
151. Noguchi K, Kitanaka C, Yamana H et al. Regulation of c-Myc through phosphorylation at Ser-62 and Ser-71 by c-Jun N-terminal kinase. J Biol Chem 1999; 274:32580-32587.
152. Whitfield J, Neame SJ, Paquet L et al. Dominant-negative c-Jun promotes neuronal survival by reducing BIM expression and inhibiting mitochondrial cytochrome c release. Neuron 2001; 29:629-643.
153. Lei K, Davis RJ. JNK phosphorylation of Bim-related members of the Bcl2 family induces Bax-dependent apoptosis. Proc Natl Acad Sci USA 2003; 100:2432-2437.
154. Putcha GV, Le S, Frank S et al. JNK-mediated BIM phosphorylation potentiates BAX-dependent apoptosis. Neuron 2003; 38:899-914.
155. Faris M, Latinis KM, Kempiak SJ et al. Stress-induced Fas ligand expression in T cells is mediated through a MEK kinase 1-regulated response element in the Fas ligand promoter. Mol Cell Biol 1998; 18:5414-5424.
156. Kasibhatla S, Brunner T, Genestier L et al. DNA damaging agents induce expression of Fas ligand and subsequent apoptosis in T lymphocytes via the activation of NF-kappa B and AP-1. Mol Cell 1998; 1:543-551.
157. Tournier C, Hess P, Yang DD et al. Requirement of JNK for stress-induced activation of the cytochrome c-mediated death pathway. Science 2000; 288:870-874.
158. De Smaele E, Zazzeroni F, Papa S et al. Induction of gadd45beta by NF-kappaB downregulates pro-apoptotic JNK signalling. Nature 2001; 414:308-313.
159. Tang G, Minemoto Y, Dibling B et al. Inhibition of JNK activation through NF-kappaB target genes. Nature 2001; 414:313-317.
160. Cardone MH, Salvesen GS, Widmann C et al. The regulation of anoikis: MEKK-1 activation requires cleavage by caspases. Cell 1997; 90:315-323.
161. Widmann C, Johnson NL, Gardner AM et al. Potentiation of apoptosis by low dose stress stimuli in cells expressing activated MEK kinase 1. Oncogene 1997; 15:2439-2447.
162. Widmann C, Gerwins P, Johnson NL et al. MEK kinase 1, a substrate for DEVD-directed caspases, is involved in genotoxin-induced apoptosis. Mol Cell Biol 1998; 18:2416-2429.
163. Low W, Smith A, Ashworth A et al. JNK activation is not required for Fas-mediated apoptosis. Oncogene 1999; 18:3737-3741.

164. Wilson DJ, Alessandrini A, Budd RC. MEK1 activation rescues Jurkat T cells from Fas-induced apoptosis. Cell Immunol 1999; 194:67-77.

165. Juo P, Kuo CJ, Reynolds SE et al. Fas activation of the p38 mitogen-activated protein kinase signalling pathway requires ICE/CED-3 family proteases. Mol Cell Biol 1997; 17:24-35.

166. Lenczowski JM, Dominguez L, Eder AM et al. Lack of a role for Jun kinase and AP-1 in Fas-induced apoptosis. Mol Cell Biol 1997; 17:170-181.

167. Rochat-Steiner V, Becker K, Micheau O et al. FIST/HIPK3: A Fas/FADD-interacting serine/threonine kinase that induces FADD phosphorylation and inhibits fas-mediated Jun NH(2)-terminal kinase activation. J Exp Med 2000; 192:1165-1174.

168. Sabapathy K, Hu Y, Kallunki T et al. JNK2 is required for efficient T-cell activation and apoptosis but not for normal lymphocyte development. Curr Biol 1999; 9:116-125.

169. Rudel T, Zenke FT, Chuang TH et al. p21-activated kinase (PAK) is required for Fas-induced JNK activation in Jurkat cells. J Immunol 1998; 160:7-11.

170. Toyoshima F, Moriguchi T, Nishida E. Fas induces cytoplasmic apoptotic responses and activation of the MKK7-JNK/SAPK and MKK6-p38 pathways independent of CPP32-like proteases. J Cell Biol 1997; 139:1005-1015.

171. Cahill MA, Peter ME, Kischkel FC et al. CD95 (APO-1/Fas) induces activation of SAP kinases downstream of ICE-like proteases. Oncogene 1996; 13:2087-2096.

172. Herr I, Wilhelm D, Meyer E et al. JNK/SAPK activity contributes to TRAIL-induced apoptosis. Cell Death Differ 1999; 6:130-135.

173. Kennedy NJ, Kataoka T, Tschopp J et al. Caspase activation is required for T cell proliferation. J Exp Med 1999; 190:1891-1896.

174. Yang X, Khosravi-Far R, Chang HY et al. Daxx, a novel Fas-binding protein that activates JNK and apoptosis. Cell 1997; 89:1067-1076.

175. Chang HY, Nishitoh H, Yang X et al. Activation of apoptosis signal-regulating kinase 1 (ASK1) by the adapter protein Daxx. Science 1998; 281:1860-1863.

176. Li H, Leo C, Zhu J et al. Sequestration and inhibition of Daxx-mediated transcriptional repression by PML. Mol Cell Biol 2000; 20:1784-1796.

177. Torii S, Egan DA, Evans RA et al. Human Daxx regulates Fas-induced apoptosis from nuclear PML oncogenic domains (PODs). EMBO J 1999; 18:6037-6049.

178. Hollenbach AD, Sublett JE, McPherson CJ et al. The Pax3-FKHR oncoprotein is unresponsive to the Pax3-associated repressor hDaxx. EMBO J 1999; 18:3702-3711.

179. Zhong S, Salomoni P, Ronchetti S et al. Promyelocytic leukemia protein (PML) and Daxx participate in a novel nuclear pathway for apoptosis. J Exp Med 2000; 191:631-640.

180. Pluta AF, Earnshaw WC, Goldberg IG. Interphase-specific association of intrinsic centromere protein CENP-C with HDaxx, a death domain-binding protein implicated in Fas-mediated cell death. J Cell Sci 1998; 111:2029-2041.

181. Chang HY, Yang X, Baltimore D. Dissecting Fas signaling with an altered-specificity death-domain mutant: Requirement of FADD binding for apoptosis but not Jun N-terminal kinase activation. Proc Natl Acad Sci USA 1999; 96:1252-1256.

182. Charette SJ, Lavoie JN, Lambert H et al. Inhibition of Daxx-mediated apoptosis by heat shock protein 27. Mol Cell Biol 2000; 20:7602-7612.

183. Ichijo H, Nishida E, Irie K et al. Induction of apoptosis by ASK1, a mammalian MAPKKK that activates SAPK/JNK and p38 signaling pathways. Science 1997; 275:90-94.

184. Hatai T, Matsuzawa A, Inoshita S et al. Execution of apoptosis signal-regulating kinase 1 (ASK1)-induced apoptosis by the mitochondria-dependent caspase activation. J Biol Chem 2000; 275:26576-26581.

185. Wollert KC, Heineke J, Westermann J et al. The cardiac Fas (APO-1/CD95) Receptor/Fas ligand system: Relation to diastolic wall stress in volume-overload hypertrophy in vivo and activation of the transcription factor AP-1 in cardiac myocytes. Circulation 2000; 101:1172-1178.

186. Badorff C, Ruetten H, Mueller S et al. Fas receptor signaling inhibits glycogen synthase kinase 3 beta and induces cardiac hypertrophy following pressure overload. J Clin Invest 2002; 109:373-381.

187. Hunter JJ, Chien KR. Signaling pathways for cardiac hypertrophy and failure. N Engl J Med 1999; 341:1276-1283.

188. Choukroun G, Hajjar R, Fry S et al. Regulation of cardiac hypertrophy in vivo by the stress-activated protein kinases/c-Jun NH(2)-terminal kinases. J Clin Invest 1999; 104:391-398.
189. Esposito G, Prasad SV, Rapacciuolo A et al. Cardiac overexpression of a G(q) inhibitor blocks induction of extracellular signal-regulated kinase and c-Jun NH(2)-terminal kinase activity in in vivo pressure overload. Circulation 2001; 103:1453-1458.
190. Cross DA, Alessi DR, Cohen P et al. Inhibition of glycogen synthase kinase-3 by insulin mediated by protein kinase B. Nature 1995; 378:785-789.
191. Haq S, Choukroun G, Kang ZB et al. Glycogen synthase kinase-3beta is a negative regulator of cardiomyocyte hypertrophy. J Cell Biol 2000; 151:117-130.
192. Alderson MR, Armitage RJ, Maraskovsky E et al. Fas transduces activation signals in normal human T lymphocytes. J Exp Med 1993; 178:2231-2235.
193. Kurasawa K, Hashimoto Y, Kasai M et al. The fas antigen is involved in thymic T-cell development as a costimulatory molecule, but not in the deletion of neglected thymocytes. J Allergy Clin Immunol 2000; 106:19-31.
194. Aggarwal BB, Singh S, LaPushin R et al. Fas antigen signals proliferation of normal human diploid fibroblast and its mechanism is different from tumor necrosis factor receptor. FEBS Lett 1995; 364:5-8.
195. Freiberg RA, Spencer DM, Choate KA et al. Fas signal transduction triggers either proliferation or apoptosis in human fibroblasts. J Invest Dermatol 1997; 108:215-219.
196. Jelaska A, Korn JH. Anti-Fas induces apoptosis and proliferation in human dermal fibroblasts: Differences between foreskin and adult fibroblasts. J Cell Physiol 1998; 175:19-29.
197. Ahn JH, Park SM, Cho HS et al. Nonapoptotic signaling pathways activated by soluble Fas ligand in serum-starved human fibroblasts. Mitogen-activated protein kinases and NF-kappaB-dependent gene expression. J Biol Chem 2001; 276:47100-47106.
198. Alam A, Cohen LY, Aouad S et al. Early activation of caspases during T lymphocyte stimulation results in selective substrate cleavage in nonapoptotic cells. J Exp Med 1999; 190:1879-1890.
199. Miossec C, Dutilleul V, Fassy F et al. Evidence for CPP32 activation in the absence of apoptosis during T lymphocyte stimulation. J Biol Chem 1997; 272:13459-13462.
200. Wilhelm S, Wagner H, Hacker G. Activation of caspase-3-like enzymes in nonapoptotic T cells. Eur J Immunol 1998; 28:891-900.
201. Pinkoski MJ, Green DR. Fas ligand, death gene. Cell Death Differ 1999; 6:1174-1181.
202. Newton K, Harris AW, Bath ML et al. A dominant interfering mutant of FADD/MORT1 enhances deletion of autoreactive thymocytes and inhibits proliferation of mature T lymphocytes. EMBO J 1998; 17:706-718.
203. Walsh CM, Wen BG, Chinnaiyan AM et al. A role for FADD in T cell activation and development. Immunity 1998; 8:439-449.
204. Zhang J, Cado D, Chen A et al. Fas-mediated apoptosis and activation-induced T-cell proliferation are defective in mice lacking FADD/Mort1. Nature 1998; 392:296-300.
205. Watanabe-Fukunaga R, Brannan CI, Copeland NG et al. Lymphoproliferation disorder in mice explained by defects in Fas antigen that mediates apoptosis. Nature 1992; 356:314-317.
206. Ramsdell F, Seaman MS, Miller RE et al. gld/gld mice are unable to express a functional ligand for Fas. Eur J Immunol 1994; 24:928-933.
207. Hackam AS, Yassa AS, Singaraja R et al. Huntingtin interacting protein 1 induces apoptosis via a novel caspase-dependent death effector domain. J Biol Chem 2000; 275:41299-41308.
208. Gervais FG, Singaraja R, Xanthoudakis S et al. Recruitment and activation of caspase-8 by the huntingtin-interacting protein Hip-1 and a novel partner Hippi. Nat Cell Biol 2002; 4:95-105.
209. Breckenridge DG, Nguyen M, Kuppig S et al. The procaspase-8 isoform, procaspase-8L, recruited to the BAP31 complex at the endoplasmic reticulum. Proc Natl Acad Sci USA 2002; 99:4331-4336.
210. Petit F, Corbeil J, Lelievre JD et al. Role of CD95-activated caspase-1 processing of IL-1beta in TCR-mediated proliferation of HIV-infected CD4(+) T cells. Eur J Immunol 2001; 31:3513-3524.
211. Desbarats J, Newell MK. Fas engagement accelerates liver regeneration after partial hepatectomy. Nat Med 2000; 6:920-923.
212. Suzuki I, Fink PJ. Maximal proliferation of cytotoxic T lymphocytes requires reverse signaling through Fas ligand. J Exp Med 1998; 187:123-128.

213. Suzuki I, Martin S, Boursalian TE et al. Fas ligand costimulates the in vivo proliferation of CD8+ T cells. J Immunol 2000; 165:5537-5543.

214. Suzuki I, Fink PJ. The dual functions of fas ligand in the regulation of peripheral CD8+ and CD4+ T cells. Proc Natl Acad Sci USA 2000; 97:1707-1712.

215. Desbarats J, Duke RC, Newell MK. Newly discovered role for Fas ligand in the cell-cycle arrest of CD4+ T cells. Nat Med 1998; 4:1377-1382.

216. Green DR, Ferguson TA. The role of Fas ligand in immune privilege. Nat Rev Mol Cell Biol 2001; 2:917-924.

217. Lau HT, Yu M, Fontana A et al. Prevention of islet allograft rejection with engineered myoblasts expressing FasL in mice. Science 1996; 273:109-112.

218. Kang SM, Schneider DB, Lin Z et al. Fas ligand expression in islets of Langerhans does not confer immune privilege and instead targets them for rapid destruction. Nat Med 1997; 3:738-743.

219. Allison J, Georgiou HM, Strasser A et al. Transgenic expression of CD95 ligand on islet beta cells induces a granulocytic infiltration but does not confer immune privilege upon islet allografts. Proc Natl Acad Sci USA 1997; 94:3943-3947.

220. Seino K, Kayagaki N, Okumura K et al. Antitumor effect of locally produced CD95 ligand. Nat Med 1997; 3:165-170.

221. Arai H, Gordon D, Nabel EG et al. Gene transfer of Fas ligand induces tumor regression in vivo. Proc Natl Acad Sci USA 1997; 94:13862-13867.

222. Shimizu M, Fontana A, Takeda Y et al. Induction of antitumor immunity with Fas/APO-1 ligand (CD95L)-transfected neuroblastoma neuro-2a cells. J Immunol 1999; 162:7350-7357.

223. Takeuchi T, Ueki T, Nishimatsu H et al. Accelerated rejection of Fas ligand-expressing heart grafts. J Immunol 1999; 162:518-522.

224. Seino K, Iwabuchi K, Kayagaki N et al. Chemotactic activity of soluble Fas ligand against phago-cytes. J Immunol 1998; 161:4484-4488.

225. Ottonello L, Tortolina G, Amelotti M et al. Soluble Fas ligand is chemotactic for human neutro-philic polymorphonuclear leukocytes. J Immunol 1999; 162:3601-3606.

226. Waku T, Fujiwara T, Shao J et al. Contribution of CD95 ligand-induced neutrophil infiltration to the bystander effect in p53 gene therapy for human cancer. J Immunol 2000; 165:5884-5890.

227. Roth W, Isenmann S, Nakamura M et al. Soluble decoy receptor 3 is expressed by malignant gliomas and suppresses CD95 ligand-induced apoptosis and chemotaxis. Cancer Res 2001; 61:2759-2765.

228. Behrens CK, Igney FH, Arnold B et al. CD95 ligand-expressing tumors are rejected in anti-tumor TCR transgenic perforin knockout mice. J Immunol 2001; 166:3240-3247.

229. Hohlbaum AM, Moe S, Marshak-Rothstein A. Opposing effects of transmembrane and soluble Fas ligand expression on inflammation and tumor cell survival. J Exp Med 2000; 191:1209-1220.

230. Choi C, Benveniste EN. Fas ligand/Fas system in the brain: Regulator of immune and apoptotic responses. Brain Res Brain Res Rev 2004; 44:65-81.

231. Shinohara H, Yagita H, Ikawa Y et al. Fas drives cell cycle progression in glioma cells via extracel-lular signal-regulated kinase activation. Cancer Res 2000; 60:1766-1772.

232. Desbarats J, Birge RB, Mimouni-Rongy M et al. Fas engagement induces neurite growth through ERK activation and p35 upregulation. Nat Cell Biol 2003; 5:118-125.

233. Kataoka T, Budd RC, Holler N et al. The caspase-8 inhibitor FLIP promotes activation of NF-kappaB and Erk signaling pathways. Curr Biol 2000; 10:640-648.

234. Rescigno M, Piguet V, Valzasina B et al. Fas engagement induces the maturation of dendritic cells (DCs), the release of interleukin (IL)-1beta, and the production of interferon gamma in the ab-sence of IL-12 during DC-T cell cognate interaction: A new role for Fas ligand in inflammatory responses. J Exp Med 2000; 192:1661-1668.

CHAPTER 7

Retrograde Fas Ligand Signaling

Andreas Linkermann, Jing Qian and Ottmar Janssen

Abstract

A s highlighted in the previous chapters, the interaction of Fas with Fas Ligand (FasL) affects many different aspects related to activation and apoptosis of Fas-expressing immune and tumor cells. Over the past five years we have learned, however, that FasL also acts as a costimulatory or accessory molecule for T cell activation. In the following chapter, we summarize what is known about FasL as a modulator of thymocyte development and selection or a regulator of mature T cell activation and effector function. Since to date almost no data are available on how a putative T cell receptor (TCR) /CD3-FasL crosstalk biochemically might work, we discuss ideas and hypotheses about the orchestration of retrograde signaling events also in the context of what is known from other members of the TNF family.

Fas Ligand as a Costimulatory Molecule

The first evidence for a reverse signaling capacity of FasL in murine T cells was published in 1998 by Suzuki and Fink.[1] They observed an inhibited alloantigen-specific proliferation of CD8[+] T cells in FasL-defective (B6-*gld*) mice compared to FasL wild-type (B6-*wt*) controls. The inhibition was not apparent when optimal doses of stimulatory anti-CD3 antibodies were applied in vitro indicating a threshold-sensitive T cell receptor (TCR)/CD3-to-FasL crosstalk. Using suboptimal doses of anti-CD3, stimulation with plate-bound but not soluble FasIgG (FasFc) augmented the proliferative signals in FasL-positive, but not FasL-negative cytotoxic T lymphocytes (CTLs). For CD8[+] T cells, these results suggested a 'positive' costimulatory effect through FasL. Interestingly, CD4[+] T cells obviously behave differently and seem to be inhibited rather than activated by FasL stimulation. Consequently, CD4[+] T cells derived from B6-*gld* mice (mutated FasL), but not B6-*lpr* mice (mutated Fas) or B6-*wt* mice proliferated vigorously upon alloantigen stimulation. Under similar conditions, the proliferative index of CD8[+] T cells of *gld* mice was only about 40% compared to wild type animals.

Support for the FasL reverse signal hypothesis was provided by Desbarats and colleagues who showed that FasL engagement in vivo selectively prevented the potent CD4[+]-cell driven inflammatory response following superantigen administration.[2] Again, the response of CD8[+] T cells obviously differed since expansion of CD8[+] T cells in these experiments was not affected. The inhibition of proliferation of CD4[+] (but not CD8[+]) T cells was associated with a FasL-mediated cell-cycle arrest followed by an increase in cell death. In addition, it was reported that IL-2 secretion in vitro was dramatically reduced by FasL stimulation with Fas-Fc fusion proteins when low doses of anti-CD3 were used. Thus, the consequence of

Fas Signaling, edited by Harald Wajant. ©2006 Landes Bioscience
and Springer Science+Business Media.

Table 1. Time course of death signaling, costimulation and FasL expression in naïve and long-term CD4⁺ and CD8⁺ T cells

	Death Signal through FasL	Costimulation through FasL	FasL Expression
Naïve CD4⁺ T cells	yes	yes	up until next AG contact
Long-term CD4⁺ T cells	yes	no	up until next AG contact
Naïve CD8⁺ T cells	no	yes	decreases beyond day 3
Long-term CD8⁺ T cells	no	yes	decreases beyond day 3

Table adapted from ref. 3.

FasL engagement on CD4⁺ T cells might be considered a potent, subset-specific, anti-inflammatory reaction.

The fact that the *gld*-mutation differently interferes with activation pathways of CD4⁺ and CD8⁺ T cells, suggests that a single molecule might play opposing roles in these two subpopulations. This phenomenon is likely to be due to either different interference with the TCR-signal or might be explained by a differential time course within immune responses of both CD4⁺ and CD8⁺ T cells. For example, it has been noted that naïve CD4⁺ T cells are responsive to FasL-mediated costimulation when Fas-mediated death is prevented.[1] Thus, the reverse FasL pathway is active in both naïve and antigen-matured cells, although it does not result in clonal expansion of CD4⁺ T cells when Fas-induced cell death is not inhibited (Table 1). In contrast, FasL mediated killing occurs late in an immune response when FasL molecules appear at high levels on the surface of CD4⁺, but not on CD8⁺ T cells which in turn become less sensitive to FasL-ligation.[1] The unique capacity of FasL to both positively and negatively regulate the peripheral T cell compartment has been referred to as the "dual functions" of FasL.[3]

In all studies published so far, FasL 'positive' reverse signaling seems to require a concomitant TCR engagement. Thus, FasL costimulation could not be demonstrated with stimuli that bypass the TCR/CD3-complex such as phorbolester and calcium ionophore. In addition, maximal antigen-induced accumulation of transferred TCR-transgenic CD8⁺ T cells requires functional Fas expression by the adoptive hosts. Moreover, in a primary immune response, adoptively transferred FasL⁺ CD8⁺ T cells proliferate better than their FasL⁻ counterparts in vivo.[4] Also, the FasL-costimulation was shown to enhance the burst size of the responding CTLs. The efficacy of immune clearance of invading pathogens is likely to depend on the burst size of CTLs and stresses the importance of FasL costimulation in this context.[4]

Fas Ligand as an Accessory Molecule in T Cell Development

The thymus is among the few organs in the body that coexpress Fas and FasL.[5;6] However, the interaction of these two molecules does not seem to be required for negative selection of superantigen- or conventional antigen-responsive thymocytes at physiological antigen levels.[7-9] With higher antigen concentration, however, FasL seems to become increasingly important in the complex orchestration of negative selection.[10] Apart from the main players in T cell maturation, e.g., TCRαβ and self-MHC/peptide complexes, an increasing number of cofactors including CD30, CD28, CD43 and CD5 influence the efficiency of β-selection and negative selection (for a review, see ref. 11). During positive selection, the modulation by costimulators or adhesion molecules still remains uncharted, although the apparent requirement for sustained TCR signaling during this checkpoint is compatible with their involvement.[12] In this scenario, Boursalian and Fink recently described Fas Ligand as an accessory molecule during positive selection.[13] They reported that in some (but not all) strains of TCR transgenic mice on a *gld*-background, T cell development is severely impaired. However, only positive selection seems to be altered whereas negative selection and death by neglect are not affected. The observations made in those TCR-transgenic mice suggest that FasL modulates positive selection of thymocytes coexpressing both class-I and class-II-restricted TCRs of moderate affinity for their selecting ligands. The authors assume that in thymocytes expressing TCRs with very low or very high avidity for the selecting antigen, FasL costimulation is not initiated or bypassed, respectively.[13]

The Molecular Basis for Retrograde Fas Ligand Signaling

The in vivo relevance of a retrograde Fas Ligand signaling is evident from the various mutant mouse systems that have been exploited. However, until recently nothing was known about the precise biochemical protein-protein interactions underlying the intracellular pathways. Although the transduction of these signals is likely to interfere with certain branches of TCR downstream signaling pathways, it still can not be excluded that a TCR-independent FasL-initiated signal by itself results in a cellular response. It is well known that a number of immune privileged tissues (see Chapter 12) and tumor-derived cell lines constitutively express FasL on the cell surface. Although nothing has been published about reverse signaling in these cells, they might be considered candidates for a broad range of costimulatory and/or accessory effects mediated by FasL.

The search for molecules involved in reverse signaling of FasL started with the analysis of putative interactors of the intracellular N-terminal stretch of the molecule. This region consists of an almost perfect proline-helix, several tyrosine phosphorylation sites and a so-called 'double' casein kinase phosphorylation motif.[14] With regard to protein-protein interactions, the presence of an extended polyproline region indicates an interaction with proline-binding modules such as Src-homology (SH) 3 domains or WW domains. In fact, in 1995 Hahne and colleagues reported a selective interaction of the murine FasL with SH3 domains of the Src-related tyrosine kinase p59[fyn].[15] However, this interaction was thought to be important in the context of the stabilization of FasL surface expression rather than for reverse signaling (which at that time was not in the focus of the scientific interest). The same is true for three FasL associated factors (FLAF1-3) that have been identified as such in a yeast-two-hybrid screen searching for FasL interactors. Although the partial sequences of these factors have been deposited in the database several years back, no further report followed.

Our laboratory utilized a collection of SH3- and WW-domain fusion proteins for pull-down protein-protein interaction screens from KFL-9 cells stably expressing high levels of FasL or from activated human T cell blasts. In fact, SH3 domains of several other Src-related kinases including Lck, Lyn, Hck, (and Fyn), the Tec kinase Itk, the Abl kinase, adapter proteins including Grb2, Grap, Mona/Gads, Nck, p47phox and the p85 subunit of the PI3-Kinase efficiently precipitated full length FasL from both cell types. In addition, in vitro interactions with WW domains of FE65, FBP-11 and Dystrophin were observed.[16] Moreover, using the FasL cytoplamic region fused to GST for pull-down experiments from human T cell lines, PACSIN and FBP17 were identified as putative intracellular binding partners for FasL.[17] Some of the adapter proteins mentioned above contain more than one SH3-domain. However, so far there is no example that all SH3 domains of a given protein bind to FasL. Many of the intracellular adaptor molecules are known from other signaling pathways, most often involved in cell proliferation and differentiation. For example, Grb2 is well known to stimulate the MAP-Kinase pathway whereas PI3-Kinase is implicated in several pathways resulting in cellular differentiation and survival. In addition, Nck was shown to be constitutively associated with the casein kinase 1γ2, indicating a possible role in the recruitment of this kinase to the cell membrane.[18] Importantly, several of these proteins are key players of the intracellular signaling cascade initiated by TCR/CD3-stimulation (for a review about FasL interacting proteins see refs. 19, 20).

However, the molecular basis of the FasL-to-TCR-crosstalk needs to be elucidated. FasL engagement on T cell clones with soluble anti-FasL mAb or FasFc fusion protein by itself does not induce any significant changes in overall tyrosine phosphorylation, Erk1/2 phosphorylation or Akt1 phosphorylation (Qian et al, submitted). In contrast, experiments using primary T cells, T cell clones and T cell blasts clearly argue for a connection in the sense of a massive downregulation of the TCR signal by FasL prestimulation. Importantly, in all experiments, extensive and longterm exposure to FasFc fusion proteins with additional crosslinking was prerequisite to induce the inhibitory effects. Thus, proliferation of peripheral blood mononuclear cells in response to low dose anti-CD3 or superantigen is blocked under optimal conditions by up to 80%. Of note, these inhibitory effects can in part be overcome by the addition of higher concentrations of exogenous IL-2. Moreover, in PHA blasts where one also observes an inhibition of proliferation (although less prominent), a reduction of CD3-inducible overall tyrosine phosphorylation becomes apparent upon preligation with crosslinked FasFc fusion proteins or anti-FasL mAb (Qian et al, submitted). At present, studies are ongoing to define the target second messenger molecules that show decreased TCR/CD3-induced tyrosine phosphorylation upon exposure to FasL crosslinkers.

Retrograde Signaling in Other TNF-Family Members

Most ligands of the tumor necrosis factor (TNF) superfamily are transmembrane proteins that share structurally related extracellular TNF homology domains (THDs) which bind to cystein-rich domains (CRDs) of their specific receptors (of the TNF-receptor superfamily, see Chapter 1). Ligand-receptor interactions of the two families orchestrate the organization and function of various facettes of a large number of tissues, and predominantly the immune system. As mentioned, all members of the TNF-family regulate key events in cellular activation, proliferation, differentiation, cell death and survival of immune cells and other tissues.[21] Since most of the TNF family-members are membrane-bound factors and require direct cell to cell contact, it seems obvious that ligand-receptor engagement should be seen as a bi-directional corsstalk rather than a one way monologue. In fact, ligand-transmitted costimulatory functions have been reported for several members of the TNF superfamily in T cells, B cells and monocytes including (CD70/CD27 TNFSF7/TNFRSF7),[22] CD153/CD30 (TNFSF8/TNFRSF8),[23;24] CD154/CD40 (TNFSF5/TNFRSF5),[25-28] 4-1BB-L/CD137(TNFSF9/TNFRSF),[29;30] OX40L/CD134 (TNFSF4/TNFRSF4),[31] TRANCE/RANK (TNFSF11/TNFRSF11),[32] LIGHT/LIGHTR (TNFSF14/TNFRSF14)[33-35] and the death factors TNF,[36] TRAIL (TNFSF10)[37] and FasL (TNFSF6).[1-4]

Importantly, a bi-directional signaling capacity has been exclusively associated with six of the 16 TNF family members (CD27L, CD30L, CD40L, CD137L (41BB), TNFα and FasL) that contain a putative casein kinase I (CKI) substrate phosphorylation motif (-SXXS-). Since such phosphorylation sites are not present in any of the remaining TNF family members, it seems likely that reverse signaling involves a casein kinase, possibly recruited by Nck. Moreover, a receptor-triggered and CKI-dependent serine phosphorylation of the corresponding residues within membrane-bound TNF has been experimentally proven.[38] In case of FasL, the N-terminus of the molecule contains a unique 'double' CKI substrate site (amino acids 17-21 in man (-SSASS-) and 17-22 in mice (-SSATSS-)). The presence of such a double motif (-S(P)XXS-) indicates a role for an additional kinase to regulate the sensitivity of the site for CKI by phosphorylation of the first serine residue.[38] To date, no data are available that demonstrate the involvement of a kinase or phosphatase in the regulation of FasL reverse signal transduction. However, preliminary results indicate that in in vitro kinase assays, GST-FasL fusion proteins become more readily phosphorylated by exogenously added CKI in the presence of the putative CKI substrate motif (Qian et al, submitted).

Conclusions and Perspectives

Retrograde signaling through the Fas Ligand is accepted as a costimulatory and accessory tool in the immune system. We have only just begun to understand the molecular basis and the interference with the TCR- and possibly other costimulatory signals. Understanding these patterns will not only help to understand T cell immunology, but also shed light on the still controversial topic, called "tumor counterattack". If a tumor expresses FasL on its surface and is capable of inducing death in either infiltrating lymphocytes or invaded tissue, the tumor cell will almost certainly receive a retrograde signal which might eventually lead to increased proliferation of the tumor. It will therefore be interesting to study how FasL bearing tumor cell change proliferation upon Fas-Fc stimulation. Another poorly understood problem in Fas Ligand biology is the regulation of expression and the removal of the dangerous protein from the surface. Since most activated T cells express Fas on their outer membrane, a removal of FasL seems required at least for memory T-cells. Retrograde Fas Ligand signaling is likely to regulate this process.

With regard to the expression of FasL, the identification of FasL interacting proteins with a known function in the remodelling of the cytoskeleton or the transport of lysosomes opens a new aspect of FasL biology. It will be very important to exactly elucidate how the storage and induced expression of FasL is achieved in the various cell types, most importantly in T cells and Natural Killer cells.[14]

For a more comprehensive overview on molecular interactions involved in FasL biology, please refer to our recently published reviews 14, 20.

References

1. Suzuki I, Fink PJ. Maximal proliferation of cytotoxic T lymphocytes requires reverse signaling through Fas ligand. J Exp Med 1998; 187(1):123-128.
2. Desbarats J, Duke RC, Newell MK. Newly discovered role for Fas ligand in the cell-cycle arrest of CD4+ T cells. Nat Med 1998; 4(12):1377-1382.
3. Suzuki I, Fink PJ. The dual functions of fas ligand in the regulation of peripheral CD8+ and CD4+ T cells. Proc Natl Acad Sci USA 2000; 97(4):1707-1712.
4. Suzuki I, Martin S, Boursalian TE et al. Fas ligand costimulates the in vivo proliferation of CD8+ T cells. J Immunol 2000; 165(10):5537-5543.
5. Brunner T, Yoo NJ, Griffith TS et al. Regulation of CD95 ligand expression: A key element in immune regulation? Behring Inst Mitt 1996; 97:161-174.
6. French LE, Hahne M, Viard I et al. Fas and Fas ligand in embryos and adult mice: Ligand expression in several immune-privileged tissues and coexpression in adult tissues characterized by apoptotic cell turnover. J Cell Biol 1996; 133(2):335-343.
7. Adachi M, Suematsu S, Suda T et al. Enhanced and accelerated lymphoproliferation in Fas-null mice. Proc Natl Acad Sci USA 1996; 93(5):2131-2136.
8. Sidman CL, Marshall JD, Von Boehmer H. Transgenic T cell receptor interactions in the lymphoproliferative and autoimmune syndromes of lpr and gld mutant mice. Eur J Immunol 1992; 22(2):499-504.
9. Singer GG, Abbas AK. The fas antigen is involved in peripheral but not thymic deletion of T lymphocytes in T cell receptor transgenic mice. Immunity 1994; 1(5):365-371.
10. Kishimoto H, Surh CD, Sprent J. A role for Fas in negative selection of thymocytes in vivo. J Exp Med 1998; 187(9):1427-1438.
11. Sprent J, Kishimoto H. The thymus and central tolerance. Philos Trans R Soc Lond B Biol Sci 2001; 356(1409):609-616.
12. Singer A. New perspectives on a developmental dilemma: The kinetic signaling model and the importance of signal duration for the CD4/CD8 lineage decision. Curr Opin Immunol 2002; 14(2):207-215.
13. Boursalian TE, Fink PJ. Mutation in fas ligand impairs maturation of thymocytes bearing moderate affinity T cell receptors. J Exp Med 2003; 198(2):349-360.

14. Janssen O, Qian J, Linkermann A et al. CD95 ligand - death factor and costimulatory molecule? Cell Death Differ 2003; 10(11):1215-1225.
15. Hane M, Lowin B, Peitsch M et al. Interaction of peptides derived from the Fas ligand with the Fyn-SH3 domain. FEBS Lett 1995; 373(3):265-268.
16. Wenzel J, Sanzenbacher R, Ghadimi M et al. Multiple interactions of the cytosolic polyproline region of the CD95 ligand: Hints for the reverse signal transduction capacity of a death factor. FEBS Lett 2001; 509(2):255-262.
17. Ghadimi MP, Sanzenbacher R, Thiede B et al. Identification of interaction partners of the cytosolic polyproline region of CD95 ligand (CD178). FEBS Lett 2002; 519(1-3):50-58.
18. Lussier G, Larose L. A casein kinase I activity is constitutively associated with Nck. J Biol Chem 1997; 272(5):2688-2694.
19. Linkermann A, Qian J, Kabelitz D et al. The Fas Ligand as a death factor and signal transducer? Signal Transduction 2003; 3(1-2):33-46.
20. Linkermann A, Qian J, Janssen O. Slowly getting a clue on CD95 ligand biology. Biochem Pharmacol 2003; 66(8):1417-1426.
21. Locksley RM, Killeen N, Lenardo MJ. The TNF and TNF receptor superfamilies: Integrating mammalian biology. Cell 2001; 104(4):487-501.
22. Lens SM, Drillenburg P, den Drijver BF et al. Aberrant expression and reverse signalling of CD70 on malignant B cells. Br J Haematol 1999; 106(2):491-503.
23. Cerutti A, Schaffer A, Goodwin RG et al. Engagement of CD153 (CD30 ligand) by CD30+ T cells inhibits class switch DNA recombination and antibody production in human IgD+ IgM+ B cells. J Immunol 2000; 165(2):786-794.
24. Wiley SR, Goodwin RG, Smith CA. Reverse signaling via CD30 ligand. J Immunol 1996; 157(8):3635-3639.
25. van Essen D, Kikutani H, Gray D. CD40 ligand-transduced costimulation of T cells in the development of helper function. Nature 1995; 378(6557):620-623.
26. Cayabyab M, Phillips JH, Lanier LL. CD40 preferentially costimulates activation of CD4+ T lymphocytes. J Immunol 1994; 152(4):1523-1531.
27. Miyashita T, McIlraith MJ, Grammer AC et al. Bidirectional regulation of human B cell responses by CD40-CD40 ligand interactions. J Immunol 1997; 158(10):4620-4633.
28. Blair PJ, Riley JL, Harlan DM et al. CD40 ligand (CD154) triggers a short-term CD4(+) T cell activation response that results in secretion of immunomodulatory cytokines and apoptosis. J Exp Med 2000; 191(4):651-660.
29. Langstein J, Michel J, Fritsche J et al. CD137 (ILA/4-1BB), a member of the TNF receptor family, induces monocyte activation via bidirectional signaling. J Immunol 1998; 160(5):2488-2494.
30. Langstein J, Michel J, Schwarz H. CD137 induces proliferation and endomitosis in monocytes. Blood 1999; 94(9):3161-3168.
31. Stuber E, Neurath M, Calderhead D et al. Cross-linking of OX40 ligand, a member of the TNF/NGF cytokine family, induces proliferation and differentiation in murine splenic B cells. Immunity 1995; 2(5):507-521.
32. Chen NJ, Huang MW, Hsieh SL. Enhanced secretion of IFN-gamma by activated Th1 cells occurs via reverse signaling through TNF-related activation-induced cytokine. J Immunol 2001; 166(1):270-276.
33. Scheu S, Alferink J, Potzel T et al. Targeted disruption of LIGHT causes defects in costimulatory T cell activation and reveals cooperation with lymphotoxin beta in mesenteric lymph node genesis. J Exp Med 2002; 195(12):1613-1624.
34. Shaikh RB, Santee S, Granger SW et al. Constitutive expression of LIGHT on T cells leads to lymphocyte activation, inflammation, and tissue destruction. J Immunol 2001; 167(11):6330-6337.
35. Morel Y, Truneh A, Sweet RW et al. The TNF superfamily members LIGHT and CD154 (CD40 ligand) costimulate induction of dendritic cell maturation and elicit specific CTL activity. J Immunol 2001; 167(5):2479-2486.
36. Eissner G, Kirchner S, Lindner H et al. Reverse signaling through transmembrane TNF confers resistance to lipopolysaccharide in human monocytes and macrophages. J Immunol 2000; 164(12):6193-6198.
37. Chou AH, Tsai HF, Lin LL et al. Enhanced proliferation and increased IFN-gamma production in T cells by signal transduced through TNF-related apoptosis-inducing ligand. J Immunol 2001; 167(3):1347-1352.
38. Watts AD, Hunt NH, Wanigasekara Y et al. A casein kinase I motif present in the cytoplasmic domain of members of the tumour necrosis factor ligand family is implicated in 'reverse signalling'. EMBO J 1999; 18(8):2119-2126.

CHAPTER 8

FasL and Fas in Liver Homeostasis and Hepatic Injuries

Maria Eugenia Guicciardi and Gregory J. Gores

Abstract

Fas is a death receptor expressed by every cell type in the liver. Engagement of Fas with its cognate ligand, Fas ligand (FasL), initiates a signaling cascade resulting in cell death by apoptosis. Fas plays a central role in maintaining liver homeostasis by contributing to the elimination of senescent cells. Moreover, in many circumstances, Fas-mediated apoptosis ensures the elimination of virus-infected or damaged cells by FasL-expressing cytotoxic T-lymphocytes. However, dysregulation of apoptosis, either by excessive or defective apoptosis, always leads to disease pathogenesis, such as liver failure, fibrosis, and carcinogenesis. In this chapter, we review the role of the Fas/FasL system in the physiology and pathophysiology of the liver.

Introduction

Efficient removal of unwanted cells (i.e., aged or virus-infected cells) with minimal inflammatory response is crucial for the maintenance of liver health. Apoptosis, a highly regulated form of cell death, represents the best way to do so. Cell death by apoptosis can be achieved through multiple and rather complex signaling pathways, each one of which is tightly controlled by the balance between a network of proteins and their endogenous inhibitors, that prevents the engagement of the cell death machinery under normal circumstances, and yet ensures a "quick and painless" elimination of the cell when needed. Indeed, apoptotic cells are ultimately fragmented into membrane-bound, organelle-containing corpses (the so-called apoptotic bodies) which are readily engulfed by neighboring cells. In the liver, in particular, the removal of the apoptotic bodies mainly relies on Kupffer cells via the interaction between cell surface receptors and cognate ligands expressed on the membrane of apoptotic cells.

Apoptosis can be triggered by activation of either an intrinsic, mitochondria-mediated pathway or an extrinsic, death receptor-mediated pathway. Although both of them can occur in the liver, the latter seems to be by far the most common, likely as a consequence of the high level of expression of many death receptors in hepatic cells. In particular, all liver cell types constitutively express Fas (CD95/APO-1) and are susceptible to Fas-mediated apoptosis in vivo (Table 1). The devastating effects of Fas on the liver have been demonstrated in mice injected with an agonistic anti-Fas antibody (which mimic the effect of the natural ligand), which induced massive hepatocyte apoptosis, liver injury and eventual death within a few hours from the injection.[1] The importance of Fas-mediated apoptosis in this model of fulminant

Fas Signaling, edited by Harald Wajant. ©2006 Landes Bioscience and Springer Science+Business Media.

Table 1. Fas/CD95 expression in liver cells

Cell Type	References	Clinical Implications
Hepatocytes	Ni et al, Exp Cell Res 1994; 215: 332-337 Galle et al, J Exp Med 1995; 182: 1223-1230	Acute liver damage, death
Cholangiocytes	Ueno et al, Hepatology 2000; 31: 966-974	Cholangiopathy in graft versus host disease
Sinusoidal endothelial cells	Cardier et al, FASEB J 1999; 13: 1950-1960 Janin et al, Blood 2002; 99: 2940-2947	Inflammatory angiogenesis and vascular lesions during Fas-mediated cytotoxicity
Stellate cells	Saile et al, Am J Pathol 1997; 151:1265-1272	Termination of activated stellate cell proliferation after liver damage
Kupffer cells	Muschen et al, Hepatology 1998; 27:200-208	Reduced clearance of apoptotic bodies, increased liver damage

hepatitis has been further confirmed in two recent studies showing that inhibition of Fas or caspase-8 (an upstream molecule in Fas signaling) expression by siRNA techniques reduces hepatocyte apoptosis and protects mice from acute liver failure.[2,3]

Fas Regulation

It appears that a tight regulation of Fas-mediated apoptosis is essential for the proper physiology of the liver. In order to avoid unnecessary activation of the Fas signaling, Fas localization is regulated by compartmentalization inside the cell. Only small amount of Fas localizes on the plasma membrane in unstimulated cells, with the majority of it being stored in cytosolic compartments, in particular, in the Golgi complex and the trans-Golgi network.[4,5] Translocation of Fas-containing vesicles to the cell surface occurs upon stimulation, providing an effective mechanism to regulate the plasma membrane density of the death receptor, and avoid its spontaneous activation.[5,6] Fas-mediated apoptosis can also be modulated by glycosylation of the receptor,[7] as well as at the transcriptional level, by directly regulating Fas expression. A composite binding site for the transcription factor NF-κB has been described at position 295 to 286 of the Fas gene promoter, which regulates activation-dependent Fas expression in lymphocytes.[8] A p53-responsive element is also located within the first intron of the Fas gene, and cooperates with three sequences in the promoter to up-regulate Fas receptor expression during drug-induced apoptosis of leukemic and hepatocellular carcinoma cell lines.[9-11] Other mechanisms of regulation of Fas-mediated apoptosis signaling are described elsewhere in this book.

FasL Regulation

Activation of Fas-mediated apoptosis requires the binding of Fas receptor to FasL-expressing cells or soluble FasL. The soluble form of FasL is generated by the cleavage from the membrane by a metalloprotease and can act as an effector molecule at a distance from the producing cell, but its ability to induce apoptosis is strongly reduced compared to the membrane-bound form.[12] This might provide an explanation for the absence of tissue damage in diseases associated with elevated circulating levels of FasL, such as hepatitis, AIDS and several types of tumors.[13-15] FasL is mainly expressed by activated T cells and its role is crucial in maintaining peripheral T- and B-cell homeostasis. In the liver, the Fas/FasL system controls hepatocyte homeostasis

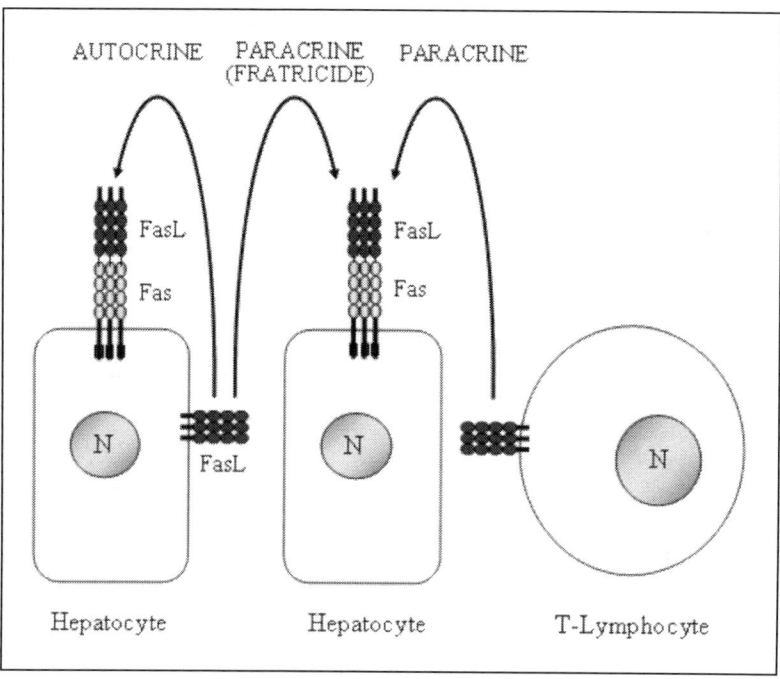

Figure 1. Schematic model of Fas-induced cell death by autocrine/paracrine mechanisms. Fas-mediated cell death can be triggered by engagement of Fas receptor with FasL expressed on a different cell (paracrine) or expressed on the same cell (autocrine). Fratricide occurs when paracrine cell death is observed between neighboring cells of the same type.

and provides a powerful tool by which FasL-positive T effector cells eliminate harmful target cells, such as virus-infected cells or cancer cells, which usually overexpress Fas (Fig. 1).[16-18] Kupffer cells, liver-specific phagocytes, can also express FasL and induce apoptosis.[19,20] Moreover, hepatocytes themselves overexpress FasL in certain pathological conditions, such as alcoholic hepatitis and Wilson's disease, which, together with a constitutively high expression of Fas, makes them capable of inducing apoptosis via fratricide (Fig. 1).[21]

The Fas/FasL System in Liver Homeostasis and Regeneration

In a physiological setting, when apoptosis is tightly regulated, and the number of cells eliminated by apoptosis equals those generated by cell division, Fas plays a crucial role in maintaining homeostasis of the peripheral lymphoid organs, as well as the liver. Studies on *lpr* (lymphoproliferation) mice, in which a functional mutation has been inserted in the *Fas* gene, have first demonstrated that loss of Fas induces lymphoadenopathy and splenomegaly.[22,23] Surprisingly, those animals did not show any gross abnormalities in nonlymphoid, Fas-expressing organs, such as liver, heart and lungs.[24] However, small amounts of functional mRNA were still detectable in the thymus and liver of the animals, which could account for the lack of phenotype in the nonlymphoid tissue.[25] Indeed, the Fas-null mice, showing a complete absence of a functional Fas molecule, do develop, besides the other abnormalities, substantial liver hyperplasia, suggesting that low levels of Fas are sufficient to maintain liver homeostasis.[26] The liver hyperplasia becomes apparent at the age of eight weeks, and it is associated with a high number of hepatocytes containing enlarged nuclei, characteristic of senescent cells.[26]Thus, these studies strongly suggest that Fas plays a key role in maintaining liver homeostasis by contributing to the removal of aged hepatocytes. Based on these observations, higher levels of Fas expression in T-lymphocytes, as well as a shorter lifespan, may explain the more severe phenotype of the lymphonodes and spleen in Fas-deficient conditions.

Beside inducing cell death, Fas, as well as other death receptors like TNF-R1, have been recently found to be also involved in cell growth and differentiation.[27] Several studies suggest that the overexpression of endogenous inhibitors of Fas signaling, such as FLIP and Bcl-2, may switch the signal from death to growth. In particular conditions, such as after partial hepatectomy, regenerating hepatocytes express higher levels of FLIP, which render them resistant to Fas-mediated apoptosis and increase Fas-mediated cell proliferation.[28] Consistent with a role of Fas in cell proliferation, *lpr* mice show delayed liver regeneration after partial hepatectomy.[28] FLIP overexpression inhibits recruitment and activation of caspase-8 at the receptor complex, diverting the signal from the apoptotic pathway toward pathways that activate the mitogen-activated protein kinase (MAPK) extracellular signal-regulated kinase (ERK) and the nuclear transcription factor NF-κB, via association of FLIP with Raf-1/MEK1 and TRAF1/TRAF2, respectively.[29] Both ERK and NF-κB are known to induce transcription of several genes involved in cell proliferation and differentiation, such as IL-2.[28] Thus, when FLIP levels are low, such as in resting hepatocytes, Fas treatment induces caspase-8 recruitment to the DISC, resulting in its activation and, ultimately, in cell death. By contrast, when FLIP levels are high, it may be preferentially recruited to the DISC, diverting the Fas signal toward ERK and NF-κB activation, and promoting cell proliferation.

The Fas/FasL System in Liver Pathophysiology

Liver diseases are often associated with disruption of hepatocyte apoptosis. Enhanced apoptosis is detectable in viral and autoimmune hepatitis, acute hepatic failure, cholestatic diseases, and metabolic disorders, as well as in alcoholic and nonalcoholic hepatitis, and in liver damage induced by chemotherapeutic drugs. Apoptosis also occurs in transplantation-associated liver damage, both in ischemia/reperfusion injury and graft rejection. On the other hand, downregulation of apoptosis leads to other diseases associated with excessive cell growth, such as hepatocellular carcinoma. Some of the human liver diseases associated with disruption of Fas-mediated apoptosis are described in greater details in this chapter (Table 2).

Fas-Inactivating Conditions

Hepatocellular Carcinoma

Hepatocellular carcinoma is the most common primary malignancy of the liver and one of the most common cancers worldwide. Multiple factors contribute to the development of this tumor, including environmental, nutritional, and metabolic factors, as well as chronic viral infections. Among the genes that have been found to be implicated in the pathogenesis of the disease, there are genes regulating the response to DNA damage, genes involved in cell cycle control, and genes involved in growth inhibition and apoptosis, which are generally downregulated.[30]

The role of Fas in carcinogenesis is strongly supported by the evidence that Fas-defective animals show increased risk of developing tumors. Down-regulation or loss of Fas expression has been observed in several tumors, including hepatocellular carcinoma, which had originated from tissue normally expressing Fas.[31,32] Loss of Fas represents an advantageous adaptation for the cancer cell, allowing the cell to survive the attack by FasL-expressing cytotoxic T-lymphocytes and NK cells. Moreover, tumor cells have been found to often express FasL, which enables them to actively kill the immune cells and create immune privileged sites.[33-35] Consistently, a

Table 2. Involvement of the Fas/FasL system in human liver diseases

Liver Disease	References
Hepatocarcinoma	Nagao M et al, Hepatology 1999; 30:413-421
	Ito Y et al, Br J Cancer 2000; 82:1211-1217
	Lee SH et al, Hum Pathol 2001;32: 250-256
	Fukuzawa Y et al, J Gastroenterol 2001; 36:681-688
	Okano H et al, Lab Invest 2003; 83:1033-1043
Chronic Viral Hepatitis	Fiore G et al, Microbios 1999; 97:29-38
	Tagashira M et al, J Clin Immunol 2000; 20:347-353
	Ehrmann J Jr. et al, Pathol Oncol Res 2000; 6:130-135
	Pianko S et al, J Viral Hepat 2001; 8:406-413
	Ibuki N et al, Liver 2002; 22:198-204
	Tang TJ et al, J Viral Hepat 2003; 10:159-167
Acute Viral Hepatitis	Rivero M et al, J Viral Hepat 2002; 9:107-113
Fulminant Hepatic Failure	Ryo K et al, Am J Gastroenterol 2000; 95:2047-2055
Alcoholic Hepatitis	Natori S et al, J Hepatol 2001; 34:248-253
	Tagami A et al, Hepatogastroenterology 2003; 50:443-448
Autoimmune Hepatitis	Fox CK et al, Liver 2001; 21:272-279
HCV-related Fibrosis	Pianko S et al, J Viral Hepat 2001; 8:406-413
Wilson's Disease	Strand S et al, Nat Med 1998; 4:588-593
Acute Allograft Rejection	Tannapfel A et al, Transplantation 1999; 67:1079-1083
Nonalcoholic Steatohepatitis	Feldstein A et al, Gastroenterology 2003; 125:437-443

complete or partial reduction of Fas expression has been described in hepatocarcinomas, with the most significant reduction observed in poorly differentiated, advanced carcinomas and negatively correlating with patient survival.[34,36,37] Overexpression of FasL was also found in 43% of the carcinomas analyzed.[34] A therapeutic approach aimed to restore Fas expression and sensitivity to Fas-mediated apoptosis in tumor cells may therefore be proven useful in the therapy of hepatocellular carcinoma, as well as other tumors.

Several chemotherapeutic drugs induce apoptosis of tumor cells by causing DNA damage and activation of the nuclear phosphoprotein p53. How p53 initiates apoptosis, however, is still poorly understood. Several genes involved in cell death have been found to be up-regulated by p53 activation, including Fas. Treatment of different cancer cell lines in vitro with multiple chemotherapeutic agents induces accumulation of nuclear p53, up-regulation of both Fas and FasL, and increased sensitivity to Fas-mediated apoptosis.[10,11] More importantly, the same effects are mimicked by microinjection of wild-type p53 into p53-expressing cells, or by transfection of wild-type p53 into p53-null cells, demonstrating that p53 activity is essential for drug-induced Fas-mediated apoptosis. However, only Fas expression seems to be directly dependent on p53 activity, since FasL induction has been observed also in the absence of functional p53.[11] Overexpression of Fas may prepare the tumor cells to be eliminated by FasL-expressing T lymphocytes, or via autocrine and paracrine apoptosis by neighboring hepatocytes. On the contrary, downregulation of Fas can lead to drug-resistance. On the other hand, expression of FasL on the membrane of the cancer cell can result in selective elimination of antitumor T lymphocytes and creation of immune-privilege sites. Therefore, modulation of p53, Fas, FasL, as well as other component of the Fas/FasL apoptotic pathway can help to develop new chemotherapeutic strategies.

Fas-Activating Conditions

Viral Hepatitis

Infection by Hepatitis B (HBV) or C virus (HCV) is the main cause for viral hepatitis and represents a worldwide health problem. The tissue damage observed during viral hepatitis can be the result of direct cytopathic effects of the virus on the infected host cells, or, more often, the result of host immune response to viral antigen. During hepatitis B, major histocompatibility complex (MHC) class I-restricted cytotoxic T lymphocytes (CTL) recognize and kill viral antigen-expressing, HBV-infected hepatocytes to clear HBV from the liver, causing the initial liver damage. Subsequently, the influx of antigen-nonspecific inflammatory cells exacerbates the tissue damage and it is likely responsible for the formation of necroinflammatory foci. Several studies have documented that CTL kill HBsAg-positive hepatocytes by Fas-dependent apoptosis. Indeed, apoptotic bodies, once referred to as Councilman bodies or acidophilic bodies, are readily identified in the liver of patients with viral hepatitis. Moreover, the expression levels of Fas are increased in the liver of patients with chronic active hepatitis B and directly correlate with disease activity such as periportal and intralobular inflammation.[21,38,39] Fas expression can be induced either by virus-specific protein expression or by inflammatory cytokines such as interleukin-1, generated after the first immune response. Finally, FasL mRNA expression in the liver of HBV-infected patients is also induced, and colocalizes with areas of lymphocytic infiltration.[21,38] Other pathways, including the TNF-α and the perforin/granzyme system, have also been implicated in hepatocyte apoptotic processes in viral hepatitis.[17,18]

Like in hepatitis B, in chronic hepatitis C, CTL specific for hepatitis C virus directly kill HCV-infected cells and remove HCV from the liver together with the infected cells. The killing of infected cells once again relies mainly on the activation of the Fas/FasL system, as demonstrated by enhanced Fas expression and increased number of FasL-positive infiltrating mononuclear cells in the liver of hepatitis C patients.[40-42]

The role of the HBV X-gene product (HBx) in Fas-mediated hepatocyte apoptosis remains controversial. HBx may stimulate the apoptotic turnover of hepatocytes as shown in a transgenic mouse model.[43] In contrast, HBx stimulation of NF-κB or JNK pathway has also been reported to protect against Fas-induced apoptosis in liver cells.[44,45] Similarly, the expression of HCV proteins may inhibit Fas-mediated apoptosis and death in transgenic mice by repressing the release of cytochrome c from the mitochondria.[46] Therefore, hepatitis virus proteins may either sensitize hepatocyte to Fas-induced apoptosis or inhibit apoptosis as a possible mechanism helping to maintain persistent infection.

In summary, the occurrence of hepatitis seems to proceed from a noninflammatory event (apoptosis) to unspecific necro-inflammation. The inflammatory process is likely the consequence of ineffective clearance of the apoptotic bodies by neighboring phagocytes, whose phagocytic capacity is overwhelmed by the large number of dying cells. The result is the release of potentially toxic or immunogenic intracellular contents, which elicits the inflammatory response and exacerbates tissue injury. Therefore, various disease stages might be explained by different response to Fas-mediated cytotoxicity. In particular, low levels of cytotoxicity may explain ineffective viral clearance in chronic hepatitis, normal levels of cytotoxicity may explain regular clearance in acute, self-limiting hepatitis, and high levels of cytotoxicity may account for over-effective clearance in fulminant hepatitis.

Alcoholic Hepatitis

The pathogenesis of alcoholic hepatitis and alcoholic cirrhosis, a consequence of chronic alcoholic hepatitis, is poorly understood. Apoptosis has been shown to play an important role in experimental ethanol-induced liver injury.[47,48] However, the importance of apoptosis in clinical alcoholic liver diseases has only recently been demonstrated. Hepatocyte apoptosis has been observed in patients with alcoholic hepatitis, and seems to directly correlate with the disease severity, being most abundant in patients with high bilirubin and AST levels, and grade 4 steatohepatitis.[49,50]

Several mechanisms have been proposed to explain alcohol-induced hepatocyte apoptosis. Induction of cytochrome P450 2E1 and the formation of reactive oxygen intermediates and lipid peroxides appears to be one of the possible mechanisms.[51,52] Cytochrome P450 2E1 generates high levels of reactive oxygen intermediates during catalysis of substrates, which may cause mitochondrial dysfunction and release of pro-apoptotic factors such as cytochrome *c* into the cytosol, where they promote caspase activation.[53] Another mechanism associated with ethanol-induced apoptosis is the activation of death receptor pathways, especially the Fas/FasL and TNF-α/TNF-R1 signaling. Indeed, Fas and FasL has been found strongly expressed in the hepatocytes of patients with alcoholic hepatitis as compared to controls, rendering the cells more sensitive to killing by cytotoxic T-lymphocyte.[49] The expression of both FasL and Fas on hepatocytes raises the possibility that liver damage in alcoholic hepatitis may be caused also by hepatocyte apoptosis occurring in an autocrine and/or paracrine fashion. Soluble Fas and FasL were also recently found to be raised in patients with severe alcoholic hepatitis, although the sources of these mediators and their biological importance remains to be investigated.[54] The increase in Fas and FasL may be the result of TNF-α-induced activation of NF-κB, a transcription factor which can up-regulate the transcription of both these genes.[8] Indeed, TNF-α serum levels are also increased during alcoholic hepatitis, and play a crucial role in mediating hepatocyte damage.[55] Beside a direct cytotoxic effect on the hepatocyte, the TNF-α/ TNF-R1 system seems to be required also for Fas-mediated cell death, as demonstrated by the increased resistance of TNF-R1/TNF-R2 double knock-out mice to Fas-induced fulminant liver injury.[56] Thus, it appears that the activation of the TNF-α/TNF-R1 system synergizes with Fas-mediated signaling to induce hepatocyte apoptosis, and that both death receptors contribute to ethanol-mediated liver injury.

Wilson's Disease

Wilson's disease is the result of excessive copper storage in the liver parenchyma. Hepatocyte apoptosis was observed in liver sections from patients with Wilson's disease, in association with upregulation of Fas and FasL on the hepatocyte cell membrane.[57] The same observations were confirmed in a model of copper overload in vitro.[57] Simultaneous expression of Fas and FasL in the same cell may promote fratricide killing of neighboring cells as already suggested in alcoholic hepatitis. Fas signaling can be activated by oxidative stress generated by excess copper accumulated in hepatocytes.[58] Upregulation of Fas in Wilson's disease has been shown to involve the p53 tumor suppressor gene, possibly as the result of DNA damage. Treatment of hepatoma cells with copper resulted in a transient increase in p53 wild-type expression and Fas expression, likely as the result of p53 transcriptional activity.[57,59] Apoptosis induced by copper can be reduced to the same extent either by antibody direct against FasL or by caspase inhibitors, suggesting that Fas might be the only apoptotic signal involved in copper-induced apoptosis. Thus, therapies employing antibodies against Fas or FasL, as well as caspase inhibitors, can be useful in acute Wilson's disease and could reduce the need for transplantation in the acute form of this disease.

Cholestatic Liver Disease

In cholestasis, an impairment of bile flow leads to accumulation of elevated concentrations of bile acids within the hepatocytes, promoting liver injury and liver failure. Although the mechanisms of liver damage associated with cholestasis are likely complex and multifactorial, the pivotal role of bile acid-mediated hepatotoxicity in the pathogenesis of the disease is unquestionable. Indeed, hydrophobic, potentially toxic bile acids have been shown to induce hepatocyte apoptosis in vitro.[5,60-62] More importantly, hepatocyte apoptosis has also been clearly identify in animal models of cholestasis.[63]

Bile acids induce hepatocyte apoptosis mainly by a Fas-dependent, FasL-independent mechanism.[60] Indeed, hepatocyte apoptosis is decreased in *lpr* mice, which express only minimal amount of Fas, after bile duct ligation (an experimental model of extrahepatic cholestasis), whereas is unaltered in bile duct-ligated *gld* mice, expressing a functionally defective FasL[63] (Fig. 2). The mechanism by which Fas undergoes oligomerization to start the apoptotic signal independent of FasL is still unclear. It has been suggested that elevated bile acid concentrations within the hepatocyte induce Fas translocation from its intracellular locations to the plasma membrane, where the increased surface density triggers its oligomerization.[5]

Other pathways are also likely to contribute to bile acid-induced hepatocyte apoptosis, as suggested by the increase in apoptosis in the *lpr* mice under conditions of persistent cholestasis.[63] Recent studies on Fas-deficient cells have demonstrated that one of these pathways involves transcriptional induction and oligomerization of another death receptor, TRAIL-R2/DR5, suggesting bile acid cytotoxicity, in the absence of Fas expression, is mediated by TRAIL/ TRAIL-R2 signaling pathway.[62] Because both Fas and TRAIL-R2 signal through activation of caspase-8/-10 and Bid to induce apoptosis, targeted inhibition of caspases or Bid could be therapeutically useful in cholestatic liver diseases. Indeed, inhibition of Bid by antisense oligonucleotides injection have already been shown to reduce hepatocyte apoptosis and liver damage in a rodent model of extrahepatic cholestasis.[64]

Nonalcoholic Steatohepatitis (NASH)

Nonalcoholic steatohepatitis (NASH) represents a subset of nonalcoholic fatty liver disease (NAFLD), characterized by the presence of steatosis along with inflammatory activity with or without fibrosis, indicating the possibility of progressive liver disease. The pathogenesis of NASH, as it relates to tissue injury, remains poorly understood. However, recent studies showed that hepatocyte apoptosis and Fas expression are enhanced in patients with NASH, even more markedly than in alcoholic hepatitis patients.[65,66] Fas-induced hepatocyte apoptosis is mediated by caspase-8-dependent cleavage of Bid, a proapoptotic protein of the Bcl-2 family, which translocates to mitochondria and induces mitochondrial dysfunction in cooperation with other proapoptotic members of the same family.[67] The mitochondrial dysfunction results in release of cytochrome *c*, activation of effector caspases (caspase-3 and -7) and apoptosis. Consistently, liver samples from NASH patients showed enhanced Fas expression, activation of caspase-3 and -7, and apoptosis, which positively correlated with biochemical and histopathologic markers of liver injury.[65,66] Generation of reactive oxygen species has also been associated with mitochondrial dysfunction, which are also able to induce apoptosis, further exacerbating tissue injury and inflammation. Therefore, therapeutic approaches aimed to inhibit Fas-mediated apoptosis may be proven useful to reduce liver damage and prevent development of cirrhosis in NASH.

Graft versus Host Disease (GVHD)

Bile-duct injury observed in hepatic graft-versus-host disease (GVHD) is regarded as an immune-mediated injury, although its precise mechanism is unclear. However, recent studies have suggested the involvement of Fas-mediated cell death in this immune-mediated cholangiopathy. Normal rat cholangiocytes constitutively express Fas, and have been found to upregulate both Fas mRNA and protein in GVHD, becoming more sensitive to Fas-mediated apoptosis.[68] On the contrary, cholangiocytes from Fas-deficient mice (*lpr*) did not undergo apoptosis after treatment with an agonistic anti-Fas antibody.[68] Moreover, administration of

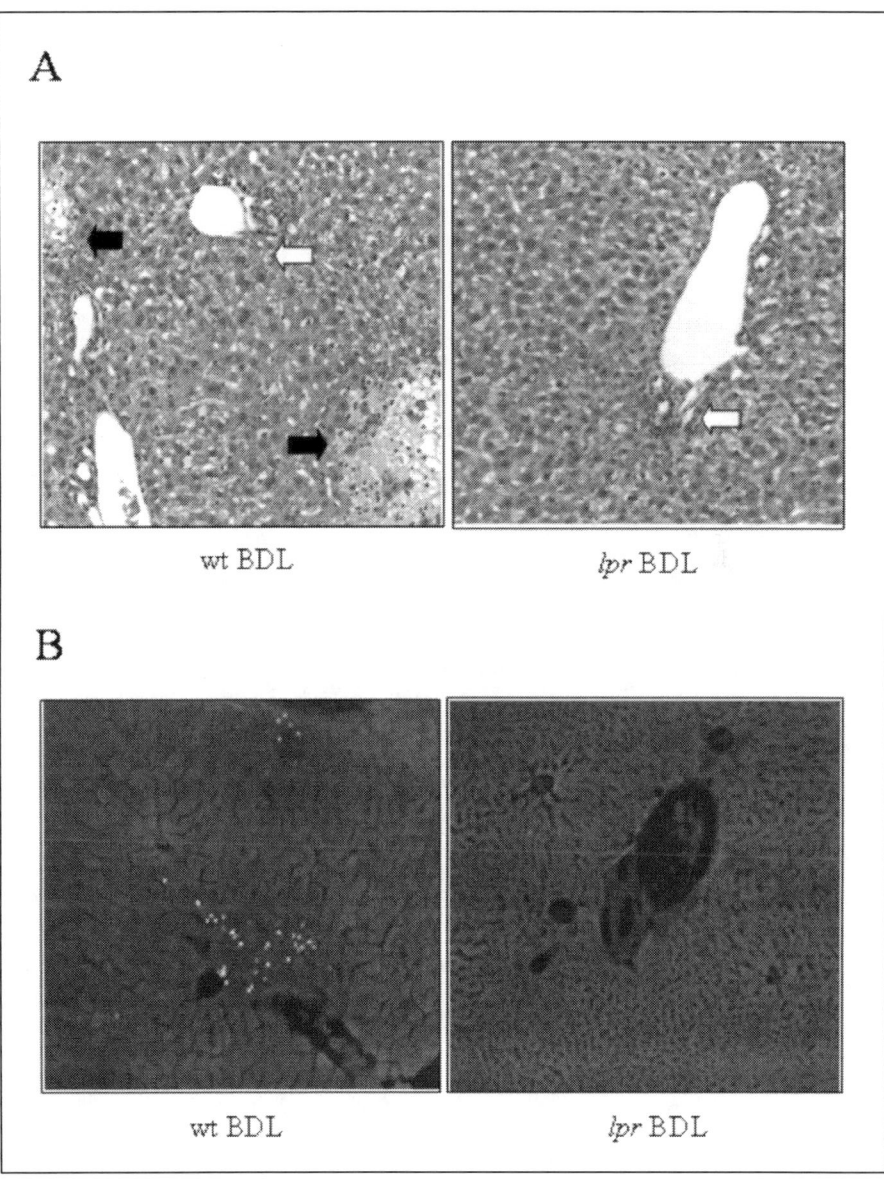

Figure 2. Role of Fas in cholestasis-induced liver damage. Histological examination of liver tissue from wild-type (wt) and Fas-deficient (*lpr*) bile duct-ligated (BDL) mice. The experimental model of extrahepatic cholestasis was carried out for 3 days, then the animals were sacrificed and the livers collected and analyzed by standard techniques. A) Histological examination of liver samples by conventional haematoxyllin/eosin (HE) staining. Extensive damage is clearly detectable in the liver of wt BDL mice, with "bile infarcts" (areas of massive hepatocyte apoptosis; black arrows), bile duct proliferation (white arrows), lymphocitic infiltration, and alteration of liver architecture. Virtually no damage is detectable in the liver of *lpr* BDL mice. B) Measurement of apoptosis by TUNEL assay. Bright green cells are TUNEL-positive and scored as apoptotic. Large clusters of apoptotic cells are detectable in the liver of wt BDL mice, but very few are present in the liver of *lpr* BDL mice.

blocking Fas-Fc fusion protein to BALB/c mice, which had been adoptively transferred with splenocytes of B10.D2 mice, has been shown to prevent the development of hepatic GVHD in vivo.[68] Therefore, these results demonstrate the involvement of Fas-mediated cell death in cholangiopathy observed in GVHD, and suggest that blocking Fas-mediated cholangiocyte apoptosis may have a therapeutic potential for hepatic GVHD.

Ischemia/Reperfusion (I/R) Damage

Ischemia/reperfusion (I/R) injury represents an unsolved problem in liver transplantation. Endothelial cell and hepatocyte apoptosis has been observed in animal models of I/R injury, which could be attenuated by using caspase inhibitors.[69,70] However, the precise mechanisms of I/R-mediated apoptosis, and the role of Fas, as well as other death receptors, in this process remain largely unknown. Fas mRNA levels have been found increased in rat livers after warm I/R by semiquantitative PCR, but the study did not investigate in which cells the overexpression occurred, or speculate on an active role of Fas in this model of apoptosis.[71] However, since pretreatment with cyclosporin A effectively reduced both I/R-induced apoptosis and Fas expression, it is conceivable that induction of Fas expression is more than just an epiphenomenon. More studies are required to clarify the role of Fas in this model of liver injury.

Fas and the Development of Liver Fibrosis and Cirrhosis

Liver fibrosis occurs during chronic liver injury as a result of excessive deposition of extracellular matrix during wound healing response. Liver fibrosis can degenerate in liver cirrhosis, characterized by the formation of thick scar tissue that alters the normal liver architecture, and can result in portal hypertension and chronic liver failure. Fibrogenesis occurs relatively late as a consequence of the activation of hepatic stellate cells and their response to the liver injury. However, the first phase of liver injury, independent of the etiology, is almost always associated with increased hepatocyte apoptosis. Recent studies have demonstrated a mechanistic link between these two pathologic events.[72,73] Apoptotic bodies generated during apoptosis are cleared from tissues by phagocytosis before they undergo autolysis and release pro-inflammatory molecules.[74] Although professional phagocytic cells such as macrophages have been the most studied as the cell type responsible for the clearance of apoptotic bodies, epithelial cells and even fibroblasts have also been shown to clear apoptotic bodies. In the liver, Kupffer cells and endothelial cells are mainly responsible for elimination of apoptotic cells. However, stellate cells, which are located in the space of Disse, adjacent to hepatocytes, are better positioned to engulf apoptotic bodies from dying hepatocytes than Kupffer cells, which are located in the intra-sinusoidal space. Although their capacity to engulf apoptotic bodies is lower than Kupffer cells, hepatic stellate cells have been shown to phagocytose apoptotic bodies in vitro.[73] More remarkably, this process is associated with increase of TGF-β1 and collagen Ia mRNA and protein, both markers of fibrogenic activity.[73] All these observations strongly suggest that phagocytosis of apoptotic bodies by stellate cells may be a potential mechanism linking liver injury to fibrosis.

The role of Fas-mediated cytoxicity in the development of liver fibrosis has been demonstrated in a model of experimental extrahepatic cholestasis. Bile duct-ligated, Fas-deficient (*lpr*) mice showed reduced expression of α-smooth muscle actin, a marker of activated stellate cells, and collagen deposition as compared to wild-type animals, pointing to a key role for Fas in the progression to liver fibrosis[72] (Fig. 3). Therefore, inhibition of Fas-mediated hepatocyte apoptosis, apoptotic bodies engulfment by stellate cells, or signaling events occurring in stellate cells as a result of phagocytosis of apoptotic bodies may be a therapeutic strategy to inhibit liver fibrogenesis.

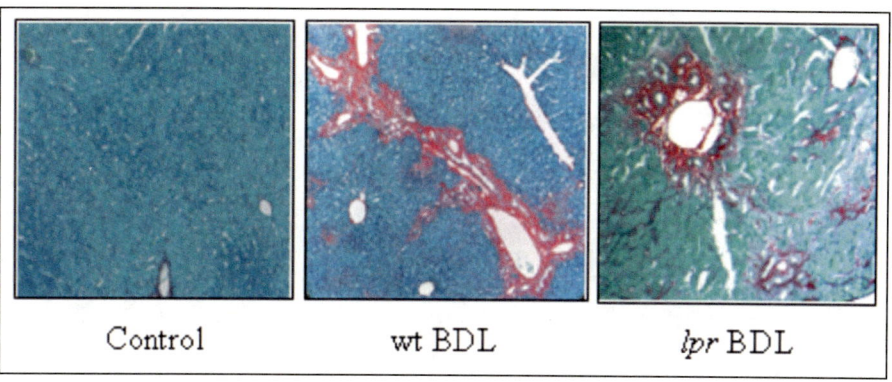

Figure 3. Role of Fas in hepatic fibrogenesis. Histological examination of liver tissue from wild-type (wt) and Fas-deficient (*lpr*) bile duct-ligated (BDL) mice. The experimental model of extrahepatic cholestasis was carried out for 3 weeks, then the animals were sacrificed, and the livers collected. Fibrosis was assessed using Sirius Red staining, a technique for detection of collagen fibers. Livers from both wt and *lpr* BDL mice show extensive bile duct proliferation. However, significant periportal and septal fibrosis was observed only in wt mice.

Conclusions

Fas plays a crucial role in maintaining liver homeostasis and liver health through regulation of cell death and survival. Its importance is demonstrated by the pathogenesis of several liver diseases where either Fas over-expression or down-regulation has been found to directly correlate with the onset of the disease and its severity. Hepatocyte apoptosis is associated with the early stage of most liver diseases, and it occurs largely through engagement of death receptors on the plasma membrane (Fig. 4). Although Fas is not the only death receptor expressed in the liver, it is so far the only one known to be express in every hepatic cell type, and this renders it the most important death receptor in hepatology. Diffuse hepatocyte apoptosis can progress into liver injury if the number of cells dying is significantly higher than the number of cells replaced by cell proliferation, and if the amount of apoptotic bodies overwhelms the clearance capacity by phagocytic cells. The subsequent autolysis of apoptotic bodies results in release of pro-inflammatory factors and recruitment of mononuclear cells, which exacerbate the inflammatory response and tissue damage. Phagocytosis of apoptotic bodies by hepatic stellate cells also triggers a pro-fibrogenic response by activating stellate cells, promoting collagen deposition in the liver parenchyma and formation of fibrotic tissue (Fig. 4).

In certain conditions, such as after partial hepatectomy, Fas can also promote cell proliferation, possibly because of higher levels of FLIP, an endogenous inhibitor of Fas-mediated apoptosis, present in regenerating hepatocytes. Thus overexpression of anti-apoptotic cell proteins (i.e., cFLIPs, cIAPs, antiapoptotic Bcl-2 family members) and activation of survival signaling pathways (i.e., NF-κB, ERK) may switch Fas signals from death to life. The ultimate fate of the cell depends upon the cell context, simultaneous signaling events, and other stimuli.

In conclusion, purposeful modulation of Fas-mediated apoptosis, either by pharmacologic or genetic manipulations, may ultimately be useful in the therapy of several human liver diseases (Fig. 4). Preliminary studies on experimental models of liver injury have already produced promising data regarding the feasibility and effectiveness of genetic inhibition of Fas as a possible therapeutic approach in fulminant liver failure.[2,3]

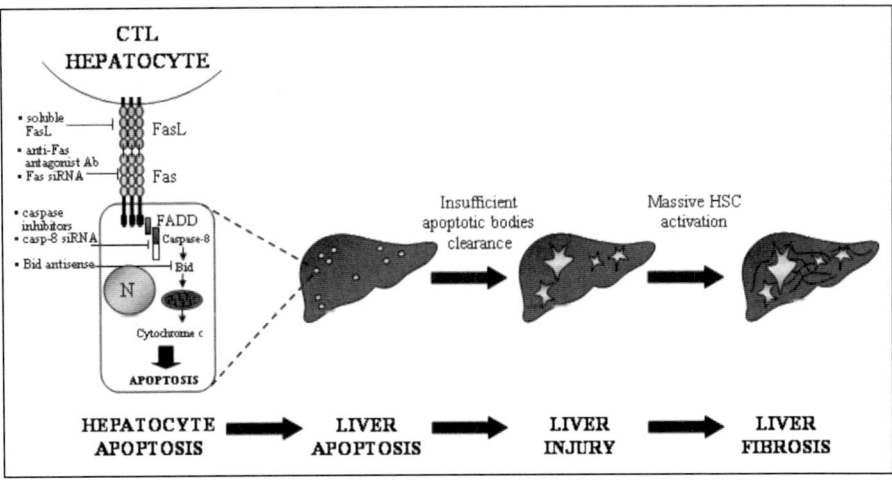

Figure 4. Schematic representation of the role of Fas in the progression of liver diseases. Fas-mediated hepatocyte apoptosis is a common early event in liver diseases. Excessive apoptosis cannot be balanced by normal cell proliferation, resulting in destruction of extensive areas of liver tissue. The damage is often exacerbated by the immune response. Moreover, phagocytosis of the resulting apoptotic bodies by hepatic stellate cells induces massive activation of stellate cells and initiates a profibrogenic process. Attenuation of Fas-mediated apoptosis by either pharmacological or genetic approaches may help slowing down, or even prevent, the progression of the liver disease.

Acknowledgements

This work has been supported by a grant from the National Institute of Health (NIH, DK41876 to G.J.G.). The authors would like to thank Dr. Ali Canbay for providing the microphotographs and Erin Bungum for her secretarial assistance.

References

1. Ogasawara J, Watanabe-Fukunaga R, Adachi M et al. Lethal effect of the anti-Fas antibody in mice. Nature 1993; 364:806-809.
2. Song E, Lee SK, Wang J et al. RNA interference targeting Fas protects mice from fulminant hepatitis. Nature Med 2003; 9:347-351.
3. Zender L, Hutker S, Liedtke C et al. Caspase-8 small interfering RNA prevents acute liver failure in mice. Proc Natl Acad Sci USA 2003; 100:7797-7802.
4. Bennet M, MacDonald K, Chan S-W et al. Cell surface trafficking of Fas: A rapid mechanism of p53-mediated apoptosis. Science 1998; 282:290-293.
5. Sodeman T, Bronk SF, Roberts PJ et al. Bile salts mediate hepatocyte apoptosis by increasing cell surface trafficking of Fas. Am J Physiol Gastrointest Liver Physiol 2000; 278:G992-G999.
6. Feng G, Kaplowitz N. Colchicine protects mice from the lethal effect of an agonistic anti- Fas antibody. J Clin Invest 2000; 105(3):329-339.
7. Peter ME, Hellbardt S, Schwartz-Albiez A et al. Cell surface sialylation plays a role in modulating sensitivity towards APO-1-mediated apoptotic cell death. Cell Death Diff 1995; 2:163-171.
8. Chan H, Bartos DP, Owen-Schaub LB. Activation-dependent transcriptional regulation of the human Fas promoter requires NF-kappaB p50-p65 recruitment. Mol Cell Biol 1999; 19(3):2098-2108.
9. Friesen C, Herr I, Krammer PH et al. Involvement of the CD95 (APO-1/Fas) receptor/ligand system in drug-induced apoptosis in leukemia cells. Nature Med 1996; 2:574-580.
10. Muller M, Strand S, Hug H et al. Drug-induced apoptosis in hepatoma cells is mediated by the CD95 (APO- 1/Fas) receptor/ligand system and involves activation of wild-type p53. J Clin Invest 1997; 99(3):403-413.

11. Muller M, Wilder S, Bannasch D et al. p53 activates the CD95 (APO-1/Fas) gene in response to DNA damage by anticancer drugs. J Exp Med 1998; 188(11):2033-2045.

12. Schneider P, Holler N, Bodmer JL et al. Conversion of membrane-bound Fas(CD95) ligand to its soluble form is associated with downregulation of its proapoptotic activity and loss of liver toxicity. J Exp Med 1998; 187(8):1205-1213.

13. Shudo K, Kinoshita K, Imamura R et al. The membrane-bound but not the soluble form of human Fas ligand is responsible for its inflammatory activity. Eur J Immunol 2001; 31(8):2504-2511.

14. Suda T, Hashimoto H, Tanaka M et al. Membrane Fas ligand kills human peripheral blood T lymphocytes, and soluble Fas ligand blocks the killing. J Exp Med 1997; 186(12):2045-2050.

15. Tanaka M, Suda T, Haze K et al. Fas ligand in human serum. Nature Med 1996; 2(3):317-322.

16. Berke G. The CTL's kiss of death. Cell 1995; 81(9-12).

17. Kagi D, Vignaux F, Ledermann B et al. Fas and perforin pathways as major mechanisms of T-cell-mediated cytotoxicity. Science 1994; 265:528-530.

18. Lowin B, Hahne M, Mattmann C et al. Cytolytic T-cell cytotoxicity is mediated through perforin and Fas lytic pathyways. Nature 1994; 370:650-652.

19. Sun Z, Wada T, Maemura K et al. Hepatic allograft-derived Kupffer cells regulate T cell response in rats. Liver Transpl 2003; 9:489-497.

20. Muschen M, Warskulat U, Peters-Regehr T et al. Involvement of CD95 (Apo-1/Fas) ligand expressed by rat Kupffer cells in hepatic immunoregulation. Gastroenterology 1999; 116:666-677.

21. Galle PR, Hofmann WJ, Walczak H et al. Involvement of the APO-1/Fas (CD95) receptor and ligand in liver damage. J Exp Med 1995; 182:1223-1230.

22. Nagata S, Golstein P. The Fas death factor. Science 1995; 267:1449-1456.

23. Nagata S, Suda T. Fas and Fas ligand: *lpr* and *gld* mutations. Immunol Today 1995; 16:39-43.

24. Cohen PL, Eisenberg RA. Lpr and gld: Single gene models of systemic autoimmunity and lymphoproliferative disease. Annu Rev Immunol 1991; 9(243-269).

25. Mariani SM, Matiba B, Armandola EA et al. The APO-1/Fas (CD95) receptor is expressed in homozygous MRL/lpr mice. Eur J Immunol 1994; 24:3119-3123.

26. Adachi M, Suematsu S, Kondo T et al. Targeted mutation in the Fas gene causes hyperplasia in peripheral lymphoid organs and in the liver. Nat Genet 1995; 11:294-300.

27. Budd RC. Death receptors couple to both cell proliferation and apoptosis. J Clin Invest 2002; 109:437-441.

28. Desbarats J, Newell MK. Fas engagement accelerates liver regeneration after partial hepatectomy. Nature Med 2000; 6:920-923.

29. Kataoka T, Budd RC, Holler N et al. The caspase-8 inhibitor FLIP promotes activation of NF-kappaB and Erk signaling pathways. Curr Biol 2000; 10:640-648.

30. Rocken C, Carl-McGrath S. Pathology and pathogenesis of hepatocellular carcinoma. Dig Dis 2001; 19:269-278.

31. Leithauser F, Dhein J, Mechtersheimer G et al. Constitutive and induced expression of APO-1, a new member of the NGF/TNF receptor superfamily, in normal and neoplastic cells. Lab Invest 1993; 69:415-429.

32. Higaki K, Yano H, Kojiro M. Fas antigen expression and its relationship with apoptosis in human hepatocellular carcinoma and noncancerous tissues. Am J Pathol 1996; 149:429-437.

33. Hahne M, Rimoldi D, Schroter M et al. Melanoma cell expression of Fas(Apo-1/CD95) ligand: Implications for tumor immune escape. Science 1996; 274(5291):1363-1366.

34. Strand S, Hofmann WJ, Hug H et al. Lymphocyte apoptosis induced by CD95 (APO-1/Fas) ligand-expressing tumor cells—a mechanism of immune evasion? Nature Med 1996; 2(12):1361-1366.

35. Griffith TS, Brunner T, Fletcher SM et al. Fas ligand-induced apoptosis as a mechanism of immune privilege. Science 1995; 17:1189-1192.

36. Ito Y, Takeda T, Umeshita K et al. Fas antigen expression in hepatocellular carcinoma tissues. Oncol Rep 1998; 5:41-44.

37. Nagao M, Nakajima Y, Hisanaga M et al. The alteration of Fas receptor and ligand system in hepatocellular carcinomas: How do hepatoma cells escape from the host immune surveillance in vivo? Hepatology 1999; 30:413-421.

38. Mochizuki K, Hayashi N, Hiramatsu N et al. Fas antigen expression in liver tissue of patients with chronic hepatitis B. J Hepatol 1996; 24:1-7.

39. Luo KX, Zhu YF, Zhang LX et al. In situ investigation of Fas/FasL expression in chronic hepatitis B virus infection and its related liver diseases. J Viral Hep 1997; 4:303-307.
40. Hiramatsu N, Hayashi N, Katayama K et al. Immunohistochemical detection of Fas antigen in liver tissue of patients with chronic hepatitis C. Hepatology 1994; 19:1354-1359.
41. Mita E, Hayashi N, Iio S et al. Role of Fas ligand in apoptosis induced by hepatitis C virus infection. Biochem Biophys Res Com 1994; 204:468-474.
42. Yoneyama K, Goto T, Miura K et al. The expression of Fas and Fas ligand, and the effects of interferon in chronic liver diseases with hepatitis C virus. Hepatol Res 2002; 24:327-337.
43. Terradillos O, De La Coste A, Pollicino T et al. The hepatitis B virus X protein abrogates Bcl-2-mediated protection against Fas apoptosis in the liver. Oncogene 2002; 21:377-386.
44. Pan J, Duan LX, Sun BS et al. Hepatitis B virus X protein protects against anti-Fas-mediated apoptosis in human liver cells by inducing NF-κB. J Gen Virol 2001; 82:171-182.
45. Diao J, Khine AA, Sarangi F et al. X protein of hepatitis B virus inhibits Fas-mediated apoptosis and is associated with up-regulation of the SAPK/JNK pathway. J Biol Chem 2001; 276:8328-8340.
46. Machida K, Tsukiyama-Kohara K, Seike E et al. Inhibition of cytochrome c release in Fas-mediated signaling pathway in transgenic mice induced to express hepatitis C viral proteins. J Biol Chem 2001; 276:12140-12146.
47. Goldin RD, Hunt NC, Clark J et al. Apoptotic bodies in a murine model of alcoholic liver disease. J Pathol 1993;171:73-76.
48. Benedetti A, Brunelli E, Risicate R et al. Subcellular changes and apoptosis induced by ethanol in rat liver. J Hepatol 1988; 6:137-143.
49. Natori S, Rust C, Stadheim LM et al. Hepatocyte apoptosis is a pathologic feature of human alcoholic hepatitis. J Hepatol 2001; 34:248-253.
50. Kawahara H, Matsuda Y, Takase S. Is apoptosis involved in alcoholic hepatitis? Alcohol 1994; 29:113-118.
51. Kurose I, Higuchi H, Miura S et al. Oxidative stress-mediated apoptosis of hepatocytes exposed to acute ethanol intoxication. Hepatology 1997; 25:368-378.
52. French SW, Wong K, Jui L et al. Effect of ethanol on cytochrome P450 2E1 lipid peroxidation, and serum protein adduct formation in relation to liver pathology pathogenesis. Exp Mol Pathol 1993; 58:61-75.
53. Green DR, Kroemer G. The central executioners of apoptosis: caspases or mitochondria? Trends Cell Biol 1998; 8:267-271.
54. Taieb J, Mathurin P, Poynard T et al. Raised plasma soluble Fas and Fas-ligand in alcoholic liver disease. Lancet 1998; 351:1930-1931.
55. McClain C, Hill D, Schmidt J et al. Cytokines and alcoholic liver disease. Semin Liver Dis 1993; 13:170-182.
56. Costelli P, Aoki P, Zingaro B et al. Mice lacking TNFα receptors 1 and 2 are resistant to death and fulminant liver injury induced by agonistic anti-Fas antibody. Cell Death Differ 2003; 10:997-1004.
57. Strand S, Hofmann WJ, Grambihler A et al. Hepatic failure and liver cell damage in acute Wilson's disease involve CD95 (APO-1/Fas) mediated apoptosis. Nat Med 1998; 4:588-593.
58. Aust SD, Morehouse LA, Thomas CE. Role of metals in oxygen radical reactions. J Free Radic Biol Med 1985; 1:3-25.
59. Narayanan VS, Fitch CA, Levenson CW. Tumor suppressor protein p53 mRNA and subcellular localization are altered by changes in cellular copper in human Hep G2 cells. J Nutr 2001; 131:1427-1432.
60. Faubion WA, Guicciardi ME, Miyoshi H et al. Toxic bile salts induce rodent hepatocyte apoptosis via direct activation of Fas. J Clin Invest 1999; 103(1):137-145.
61. Guicciardi ME, Gores GJ. Bile acid-mediated hepatocyte apoptosis and cholestatic liver disease. Digest Liver Dis 2002; 34:387-392.
62. Higuchi H, Bronk SF, Takikawa Y et al. The bile acid glycochenodeoxycholate induces trail-receptor 2/DR5 expression and apoptosis. J Biol Chem 2001; 276(42):38610-38618.
63. Miyoshi H, Rust C, Roberts PJ et al. Hepatocyte apoptosis after bile duct ligation in the mouse involves Fas. Gastroenterology 1999; 117(3):669-677.

64. Higuchi H, Miyoshi H, Bronk SF et al. Bid antisense attenuates bile acid-induced apoptosis and cholestatic liver injury. J Pharmacol Exp Ther 2001; 299:866-873.

65. Feldstein AE, Canbay A, Angulo P et al. Hepatocyte apoptosis and Fas expression are prominent features of human nonalcoholic steatohepatitis. Gastroenterology 2003; 125:437-443.

66. Feldstein AE, Canbay A, Guicciardi ME et al. Obesity is associated with fatty acid-induced upregulation of Fas (CD95) expression in the liver and increased sensitivity to Fas-mediated liver injury. J Hepatol 2003; in press.

67. Scaffidi C, Fulda S, Srinivasan A et al. Two CD95 (APO-1/Fas) signaling pathways. Embo J 1998; 17(6):1675-1687.

68. Ueno Y, Ishii M, Yahagi K et al. Fas-mediated cholangiopathy in the murine model of graft versus host disease. Hepatology 2000; 31:966-974.

69. Kohli V, Selzner M, Madden JF et al. Endothelial cell and hepatocyte deaths occur by apoptosis after ischemia-reperfusion injury in the rat liver. Transplantation 1999; 67(1099-1105).

70. Natori S, Selzner M, Valentino KL et al. Apoptosis of sinusoidal endothelial cells occurs during liver preservation injury by a caspase-dependent mechanism. Transplantation 1999; 68:89-96.

71. Saxton NE, Barclay JL, Clouston AD et al. Cyclosporin A pretreatment in a rat model of warm ischaemia/reperfusion injury. J Hepatol 2002; 36:241-247.

72. Canbay A, Higuchi H, Bronk SF et al. Fas enhances fibrogenesis in the bile duct ligated mouse: A link between apoptosis and fibrosis. Gastroenterology 2002; 123:1323-1330.

73. Canbay A, Taimr P, Torok N et al. Apoptotic body ingulfment by a human stellate cell line is profibrogenic. Lab Invest 2003; 83:655-663.

74. Platt N, da Silva RP, Gordon S. Recognizing death: The phagocytosis of apoptotic cells. Trends Cell Biol 1998; 8:365-372.

CHAPTER 9

Fas-Activation, Development and Homeostasis of T Cells

Georg Häcker

Abstract

Fas (APO-1/CD95) is found on various cells of the immune system where its expression depends on differentiation and activation status of the cells. Analysis of the function of Fas on T lymphocytes has been the objective of many studies. Like in most cells carrying Fas, ligation by antibodies or FasL can induce apoptosis in T cells. Over the years, Fas has been implicated in the regulation of many aspects of T cell physiology. However, more recent data have challenged some of the earlier views, and the matter of the importance of Fas for life and death of a T cell is, in some aspects, controversial. Fas has been suggested to play a role in selection/development of a T cell, the activation of a resting T cell and the homeostasis of a T cell population, an area that has attracted much attention. In this chapter I will focus on the discussion of the data available on the role of Fas in these fields. The role of FasL on activated T cells will be covered briefly.

Expression of Fas on T Cells and Susceptibility to Fas-Induced Apoptosis

T cells develop through a series of differentiation events from bone marrow-derived precursors in the thymus; these developing T cells are referred to as thymocytes. Thymocytes express high levels of Fas[1,2] and are susceptible to Fas-induced apoptosis. The regulation of Fas-expression and susceptibility appears to be somewhat more complex in peripheral T cells. Little or no Fas is expressed on resting peripheral blood T cells but its expression is rapidly induced upon mitogenic stimulation.[3,4] However, short time (for one day)-stimulated T cells are, despite the expression of Fas on their surface, not susceptible to Fas-induced apoptosis.[5] Susceptibility is acquired during prolonged culture and requires the presence of IL-2.[6] What the function is of Fas on freshly activated T cells is unclear, but it may be connected to cell activation (see below).

Lpr and Gld Mice

Two naturally occurring mutant strains of mice have contributed a large part of what we know about the role of Fas in T cells. Both strains have the unusual phenotype of 'lymphoproliferation', i.e., as they age, they accumulate lymphocytes in their lymph organs. We now know that in both strains there is an inactivating mutation in the Fas/FasL-system. The mutation *lpr* ('lymphoproliferation') is a near-complete loss of function mutation of Fas,[7]

Fas Signaling, edited by Harald Wajant. ©2006 Landes Bioscience and Springer Science+Business Media.

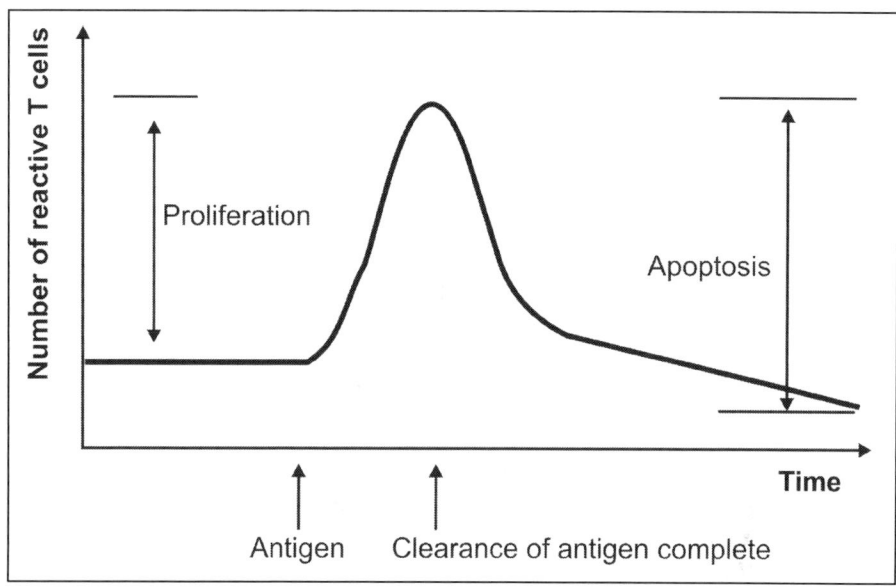

Figure 1. Cell expansion and cell death in the T cell immune response. Upon antigen exposure in vivo (infection or the injection of antigen/superantigen) the T cells with specificity for the antigen in question are activated and proliferate, thus expanding in number. Once the antigen has been cleared away, activated T cells start to die by apoptosis, and their number is reduced to about (possibly even below) the original number.

the phenotype of *gld* ('generalised lymphoproliferative disease'[8])-mice is caused by a loss of function of the FasL-gene. Mice homozygous for either mutation accumulate an unusual type of lymphocyte in the peripheral lymphatic system. These cells, CD4⁻CD8⁻ T cells (whereas normal T cells are either CD4⁺CD8⁻ or CD4⁻CD8⁺) are thought to develop from T cells that have been stimulated by antigen exposure, and there is evidence that a similar type of T cell appears when, at the end of an immune reaction, the antigen has been cleared away from the body.[9] Normally the majority of reactive T cells dies once the antigenic stimulus has disappeared (Fig. 1), and this cell death prevents the accumulation of T cells. It appears that in lpr and gld mice the 'strange T cells' are generated in the course of the immune responses going on during the life of the animals.

The same but even more pronounced phenotype has been reported for mice in which Fas has been inactivated by gene targeting[10] indicating that it is indeed a loss of a functional Fas/FasL-system that caused this effect. Mutations of the Fas/FasL-genes have also been described in human patients that, by and large, recapitulate this phenotype of reduced susceptibility to cell death, i.e., accumulation of lymphocytes and, in some cases, autoimmune disease (for review see ref. 11). The most immediate explanation for these findings would be that Fas is involved in the death of the reactive T cells at the conclusion of the immune response and that this death does not occur in the mutant mice, leading to the survival of stimulated T cells that accumulate and cause splenomegaly and lymphadenopathy. However, recent works indicates that this view is at least incomplete and, more likely, incorrect. The conflicting models will be discussed in more detail below.

Development and Selection of T Cells

T lymphocytes have to pass through a process referred to as thymic selection before they are released as mature T cells. Broadly speaking, selection in the thymus has two aspects: 'negative selection', where immature T cells bearing potentially autoreactive receptors are eliminated and 'positive selection' where thymocytes deemed to possess the right features for antigen-recognition are favoured. Both processes operate by either inducing or inhibiting cell death by apoptosis.[12] However, although massive apoptosis undoubtedly occurs in the thymus,[13] Fas appears not to be involved in this process. Lpr/lpr-mice show normal Vβ T cell receptor usage which implies that thymic selection is normal.[14] Furthermore, in an experimental model of antigen-driven negative selection in the thymus T cells were normally deleted in lpr/lpr mice, again indicating that the selecting mechanisms do not involve Fas-signalling.[15]

During T cell development there exists the chance that autoreactive T cells are generated that would attack self-tissues and cause autoimmunity. Thymic selection is the safeguard against this possibility, since autoreactive T cells are eliminated in the thymus. It may therefore be somewhat surprising that, despite apparently normal thymic selection, mice and human patients with disrupted Fas-function do develop signs of autoimmune disease. This appears, however, to be due to a 'post-thymic' problem, possibly related to a change in lymphocyte homeostasis (see below).

Involvement of Fas-Signalling in the Activation of T Cells

Members of the TNFR family of receptors not only can induce cell death but also have strong effects on gene expression and cell activation/differentiation. The spectrum of these effects is broad. At one end of the spectrum, TNFR1 is a weak inducer of apoptosis but a strong inducer of cell activation/differentiation, at the other end, Fas (and probably other death receptors) predominantly induce apoptosis. That TNF can aid proliferation in resting T cells has been described 15 years ago.[16] More recent work has shown that Fas-signalling can, surprisingly, also enhance TCR-driven proliferation in primary T cells.[17,18] This finding is most likely related to another set of unexpected data, i.e., that molecules known to be involved in the Fas-induced induction of apoptosis also play a role in the proliferation of T cells. Of these downstream molecules, the 'adapter' protein FADD/MORT1,[19-22] the somewhat enigmatic protein FLIP[23] and caspase-8[24] have all been clearly shown to contribute to the TCR-mediated induction of proliferation of resting T cells of genetically modified mice. For instance, T cells without functional FADD are much reduced in their proliferative capacity. Of note, this function appears to be limited to a proliferation defect while, at least in most models, other activation steps such as kinase activation and cytokine production were normal (for review see ref. 25). Although Fas is thus able to promote proliferation, probably by virtue of its capacity to activate caspase-8, there seems to be no evidence that Fas exerts this effect under normal circumstances. T cells from *lpr* mice (lacking functional Fas) proliferate normally, at least in vitro in response to mitogens.[19] It should be noted that proliferation of T cells from older *lpr*-mice do show a defective proliferative response.[26] However, this appears to be limited to the subset of unusual T cells that develop in these mice over time, while lpr-T cells with normal phenotype proliferate normally. Although this does not formally exclude the possibility that Fas is involved in antigen-driven T cell proliferation, a function that in the absence of Fas might be covered by other death receptors, it makes this possibility at least unlikely. The more plausible model appears to be that FADD/FLIP-dependent caspase-8-activation is required for T cell proliferation but that this activation normally is regulated independently of Fas. When, however, an artificial Fas-stimulus is applied experimentally, this enhances caspase-8-activation and will result in an enhancement of the proliferative signal (Fig. 2).

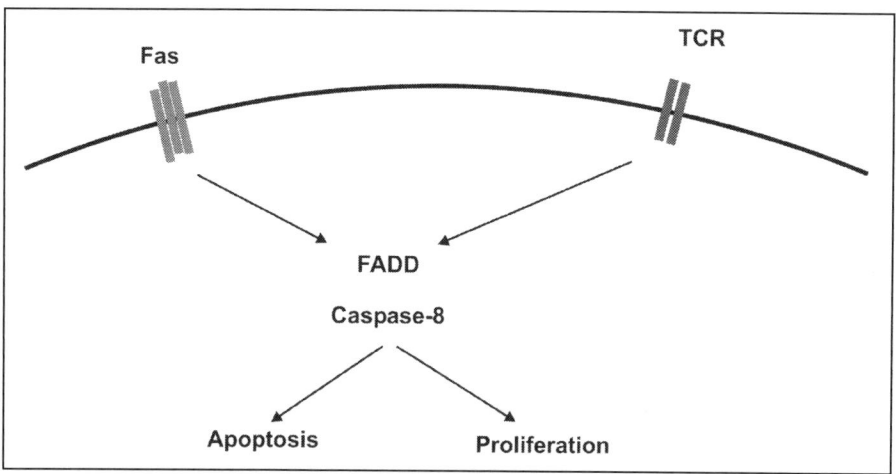

Figure 2. Fas and Fas-signalling molecules in T cell activation. The signalling molecules FADD and caspase-8 (and probably FLIP) have at least two functions, in the transduction of a death-receptor signal and in the transduction of a signal that leads to proliferation (the latter has only been shown in T cells). During normal T cell receptor (TCR) mediated cell activation, a proliferative signal is generated that involves the activity of FADD and caspase-8. A Fas-signal is best known for its capacity to induce (via FADD/caspase-8) apoptosis. A Fas-signal can, however, also enhance proliferation. On the other hand, lpr-T cells have no growth defect. The most likely explanation for this is that the caspase-8-activation that can also be achieved by Fas-triggering contributes to T cell activation, although under normal circumstances such a signal is not received.

Role of Fas for T Cell Homeostasis

There is no doubt that apoptosis plays a major role in homeostatic processes in numerous tissues: somehow, the number of cells must be kept constant in a given tissue under normal circumstances. Where proliferation drives the generation of new cells, the excess cells must disappear at some stage, and this is achieved through apoptosis. Clearly, this scenario is relevant for T cells during an immune response. Upon recognition of their antigen on antigen-presenting cells during an infection or upon vaccination, individual T cell clones develop by intense proliferation. Once the antigen has been eliminated the size of the T cell pool is reduced to a cell number near the initial figure, and this reduction occurs by apoptosis of the majority of activated T cells (Fig. 1). Mice from strains with an inactive Fas/FasL-system (*lpr, gld*) show the striking phenotype of massive accumulation of T lymphocytes over time (see above). The accumulating cells display a pattern of cell surface protein expression distinct from 'normal' peripheral T cells (they are CD4$^-$CD8$^-$B220$^+$ whereas the majority of peripheral T cells are CD4$^+$CD8$^-$B220$^-$ or CD4$^-$CD8$^+$B220$^-$). Analysis of promoter methylation suggested that these unusual T cells have developed by down-regulation of CD8,[27] and the obvious speculation is that these cells have, due to their Fas-defect, escaped apoptosis at the end of an immune response. However, recent work strongly suggests that, at least in the experimental models investigated, this form of cell death can occur independently of Fas.

Activation-Induced Cell Death

Homeostasis in T cells may involve the aspect of continuous generation of new cells from the bone marrow and the thymus but little is known about apoptosis in the regulation of this process. A large number of studies have, however, addressed the issue of T cell apoptosis at the end of an immune response. Activated T cells that are no longer required, once the antigen has gone, die by apoptosis. This form of cell death has for a number of years commonly been referred to as activation induced cell death (AICD), and a role for Fas-signalling has been suggested in this process. Nevertheless, recent studies have found that a pathway that does not require any input from Fas governs this form of cell death, and the term 'activated T cell autonomous cell death' (ACAD) has been suggested (for recent discussions of the two models see also refs. 28 and 29). With the availability of new tools, especially genetically modified mice and with the recent advances in our understanding of cell death regulation, the picture of T cell apoptosis at the end of an immune response has become clearer, and the two main models will be discussed here.

The term AICD describes cell death that occurs upon ligation of the TCR. While such ligation, by experimental stimulation with antibodies or upon recognition of MHC-peptide-complexes, results in activation of mature T cells, it can also be a strong stimulus to undergo apoptosis in some T cell populations. Whether a TCR-signal leads to cell activation or dell death is probably largely a question of signal strength. A strong signal can lead to apoptosis, a principle that has originally been described as the induction of cell lysis in T hybridomas.[30] The same phenomenon was soon after also observed in thymocytes, and the mechanism of this form of cell death (then termed AICD) was identified as apoptosis.[31] There can be no doubt that AICD can be reliably induced experimentally in vitro. This does not, however, automatically mean that AICD occurs in vivo, and the experimental support for such a role is less than cogent. Some years ago, a critical importance of Fas/FasL for AICD in vitro was reported by a number of groups.[32-36] These findings had obviously great implications for the regulation of T cell homeostasis in vivo, and the prediction was tested in a number of experimental situations. In one model, very similar results were indeed obtained in vivo: when TCR-transgenic mice were injected with the peptide that is recognised by these T cells, peripheral T cells disappeared while transgenic *lpr*-T cells did not.[15] This is probably the same effect as seen in vitro, namely that a strong stimulus through the TCR causes Fas-dependant apoptosis. But again, does this reflect the normal situation during an immune response? The scenario derived from this hypothesis would be that T cells become activated upon antigen exposure in vivo, such as during an infection, and then die when a second or a prolonged T cell-receptor signal is received, i.e., upon renewed or continued recognition of the antigen. This would appear at the very least counter-intuitive and, although as a mechanism conceivable, it would certainly appear to be beneficial for the infectious agent. The hypothesis has been tested numerous times in vivo. The model that was used by most studies addressing this question was the injection of staphylococcal enterotoxins into mice. These agents act as so-called superantigens, i.e., they stimulate T cells that carry certain Vβ-elements on their T cell receptor (for instance, T cells that express Vβ8 respond to staphylococcal enterotoxin B [SEB]). Upon injection of SEB, expansion of the Vβ8⁺ population of T cells is seen, followed by a deletion of these cells by apoptosis.[37]

When used to study the deletion of activated T cells, this model has the drawback that it operates without the adequate T cell costimulation normally provided by activated dendritic cells. Although this problem is serious, since such costimulation impacts on the survival of activated T cells,[38] the model is easy to perform, and most studies investigating cell death of activated T cells in vivo have relied on it. The question of whether Fas is required for deletion upon SEB-injection has prompted different answers. Some investigators have found that in *lpr* or *gld* mice the deletion is less pronounced[39,40] while others were unable to confirm this.[41] The finding that T cells without functional Fas still undergo apoptosis in this situation does not fully exclude the possibility that Fas partakes in this deletion. It is also conceivable that other mechanisms fill in if Fas is absent ('functional redundancy'), and it has been suggested that TNF may in this way cover for Fas-absence.[42] A recent study again addressed this issue using mice that lacked Fas (lpr-mice) crossed onto mice deficient for both TNF-receptors. Again, deletion subsequent to SEB-induced T cell expansion was normal in these triple mutant T cells.[43] If functional redundancy is the reason for the normal performance of lpr-T cells, it must thus involve molecules other than TNF-receptors. Furthermore, recent data strongly argue that a pathway that is clearly separate from Fas-induced apoptosis implements this form of cell death (see below). That SEB can induce apoptosis in *lpr* T cells has also been shown in a model of autoimmunity,[44] and deletion of reactive lpr T cells at the conclusion of an anti-viral immune response was found to be normal in these mice.[45]

The available experimental data are therefore, to a degree, controversial. However, it must thus be stated that although evidence has been presented that Fas does participate in the deletion of T cells at the end of an immune response, there is much evidence that it does not. In addition, as already mentioned, a new plausible model of a mechanism for this T cell deletion is emerging that does not require Fas. According to this model, the death of activated T cells is mediated by the activation of the so-called BH3-only protein Bim (a pro-apoptotic member of the Bcl-2 protein family), as convincingly shown in Bim-deficient mice.[43] Together with the finding that Fas-induced apoptosis proceeds normally in Bim-deficient T cells,[46] this provides a molecular model of the death (ACAD) of activated T cells in vivo that does not involve Fas. Although the discovery of an important alternative mechanism does not exclude the possibility that Fas also plays a role (and it has indeed been suggested that two pathways can complement each other),[47] it appears on the balance of the evidence available that the Fas/FasL-system is at least not critically involved in the deletion of activated peripheral T cells.

Role of Fas in the Accumulation of Unusual T Cells

As discussed above, a major contribution of Fas to the death of activated T cells in vivo is uncertain. But why, then, do *lpr* and *gld* mice as well as human patients with Fas-signalling defects accumulate lymphocytes and develop autoimmune phenomena (at least on certain genetic backgrounds)? One clue might come from a human patient with complete Fas-deficiency. In this patient, numerous mitotic figures were observed in sections from spleen and lymph node.[11] This might indicate that increased proliferation contributes to the accumulation of lymphocytes in the absence of Fas.

Genetically modified mice contribute further important evidence, in particular mice that either express in their T cells the cowpox inhibitor of caspase-8, CrmA[48] or a dominant negative form of FADD,[19,22,21] or that lack caspase-8 specifically in the T cell compartment.[24] T cells from all of these mice are protected against Fas-induced apoptosis, but none of them develop the *lpr* or *gld* phenotype. It has to be taken into account that T cells without functional FADD or caspase-8 show reduced proliferation in response to mitogenic stimuli (see above), an effect that might preclude the accumulation of T cells. But T cells from CrmA-transgenic mice proliferate normally (but fail to develop the T cells seen in lpr mice) despite the protection against Fas-induced apoptosis.[48] This suggests that lymphoproliferation and autoimmunity are phenomena that do not result from a T cell autonomous defect in Fas-induced apoptosis. Two possibilities have to be considered: either the Fas-contribution on T cells is not related to Fas-mediated killing but some other mechanism conveyed through Fas, or a Fas-dependent contribution from cells outside the T cell compartment is required. In this context it should be mentioned that dendritic cells (DC), the main antigen-presenting cells during an immune reaction, express Fas but are not susceptible to Fas-mediated killing. Furthermore, ligation of Fas in DC has been shown to transmit a maturation signal to the cell.[49] Since DC-maturation is a factor that critically determines the outcome of an immune response, it is (although speculative) conceivable that the Fas-defect on DC (or other antigen-presenting cells) rather than on the responding T cells is the cause of lymphoproliferation. However, which cell type does contribute to the lpr/gld-phenotype will only become clear once targeted deletions of Fas on individual populations have been performed.

Fas and 'Immune Privilege'

Here I will briefly discuss the evidence that the Fas/Fas-ligand system is the cause of or at least contributes to the phenomenon termed 'immune privilege'. This term has been coined to describe the finding that at certain 'privileged' sites of the body immune responses do not occur despite the presence of antigens that would, in other, 'normal' parts, trigger such a response. Immune privileged sites include brain, eye, testis, ovary, pregnant uterus and placenta. In a similar way, the term 'immune privilege' has been applied to tumours. Based on the results described below for transplanted tissue, it has been postulated that FasL-expression by tumour cells kills T cells that would otherwise reject the tumour (a mechanism referred to as 'tumour counter attack'). This aspect is covered elsewhere in this book and will here not be considered further; suffice to say that this view is not supported by much evidence (for a critical assessment see ref. 50).

The idea that the Fas/FasL system is involved in the maintenance of immune privilege was first put forward by work published in 1995. Two important findings were then reported: first, it was shown that the expression of FasL on testicular Sertoli cells prevented the rejection of mouse testes when transplanted into allogeneic recipients.[51] Second, viral infection of the eye caused invasion of immune cells and an inflammatory response in mice without a functional Fas/FasL system but not in normal mice.[52] The interpretation of these data was straightforward: FasL expressed by resident cells from immune privileged tissues (such as Sertoli cells or corneal epithelium) acts to kill invading Fas-positive immune cells (especially T cells) thus preventing inflammation and/or rejection.

The implications of these findings were far-reaching. If the principle of immune privilege was the intrinsic quality of a single molecule, as implied for FasL, it would be possible to avoid the rejection of mis-matched allografts by the expression of FasL on transplanted tissues. However, the initial euphoria was greatly dampened by a number of studies that failed either to reproduce the initial findings or to confirm the predictions of this hypothesis in different experimental settings. The most salient example is perhaps the rejection of pancreatic islet β cells engineered to express FasL;[53] this case is also remarkable for the fact that it led to the retraction of a commentary on the possibilities of the use of FasL to prevent rejection.[54]

The most extensively studied example of immune privilege in transplantation is the transfer of corneal allografts. A number of studies have reported that the Fas/FasL-system contributes to the survival of MHC noncompatible grafts (see ref. 55 for review). But even in this prime example for a role of Fas/FasL in immune privilege, the evidence is less than unequivocal. One study, for instance, found that in autoimmune uveitis the expression of Fas or FasL on eye tissue did not affect the severity of the inflammation (as would have been predicted, although in this study the Fas/FasL-system on cells from the immune system did participate[56]). While outright denials of a role for Fas/FasL-expression on tissue-resident cells in immune privilege can be found in the literature,[57] some authors still uphold this hypothesis.[55] It cannot be disputed, however, that the now repeatedly reported capacity of FasL on transplanted tissue to attract neutrophil granulocytes and cause abscess formation[53,50] is difficult to reconcile with the principle of immune privilege, which is associated with a lack of inflammation. Today, even advocates of the FasL-hypothesis emphasise that 'microenvironment' is an important issue in immune privilege,[55] and the more common view is that FasL may be a contributing factor rather than a stand-alone mediator of immune evasion in corneal immune privilege.[58]

FasL as an Effector Molecule of Cytotoxic T Cells

So far I have discussed the role of Fas on T cells. FasL is also expressed on T cells where it likely participates in a T lymphocyte effector function, i.e., killing of other cells. The effector functions of activated T cells, i.e., their effective contribution to immune defence, has two chief aspects: the secretion of cytokines (T cell help) and the lysis of dangerous cells such as virus-infected and tumour cells. Upon recognition by the T cell receptor, the T effector cell (then called cytotoxic T lymphocyte [CTL] or killer cell) kills the 'target' cell. Killing is achieved by the induction of apoptosis in the target cell, and CTL use two main mechanisms to activate the apoptotic apparatus: the secretion of cytotoxic granula containing various enzymes and the ligation of Fas on the target cell by FasL on the CTL (natural killer [NK] cells that also can induce apoptosis in target cells appear to use the TNF-related apoptosis inducing ligand [TRAIL] rather than FasL at least for tumour surveillance[59]). What the respective contributions are that these two mechanism make to target cell destruction in vivo is hard to tell, especially since they probably can replace each other in models where one is missing. It is clear that CTL can use either in in vitro systems,[60] and in at least one (however artificial) in vivo situation the presence of FasL on T cells was required to cause graft versus host disease (where transferred bone marrow cells attack tissues of the recipient[61]).

Concluding Remarks

The role of Fas on T cells has received enormous attention over the past decade. Despite the resulting wealth of experimental results, the precise role of the Fas/FasL-system in the immune system is by far not clear. Mutant mice unequivocally tell us that Fas and FasL are required for some aspects of lymphocyte physiology but the precise cellular and molecular details of this role remain uncertain. Recent work argues that earlier studies have overestimated the role of Fas on T cells but it can hardly be stated that the case was closed. We still look forward to learning more about the question on which cell and in what circumstances Fas-induced apoptosis shapes the immune system and the immune response.

References

1. Ogasawara J, Watanabe-Fukunaga R, Adachi M et al. Lethal effect of the anti-Fas antibody in mice. Nature 1993; 364:806-809.
2. Drappa J, Brot N, Elkon KB. The Fas protein is expressed at high levels on CD4+CD8+ thymocytes and activated mature lymphocytes in normal mice but not in the lupus-prone strain, MRL lpr/lpr. Proc Natl Acad Sci USA 1993; 90:10340-10344.
3. Miyawaki T, Uehara T, Nibu R et al. Differential expression of apoptosis-related Fas antigen on lymphocyte subpopulations in human peripheral blood. J Immunol 1992; 149:3753-3758.
4. Krammer PH, Dhein J, Walczak H et al. The role of APO-1-mediated apoptosis in the immune system. Immunol Rev 1994; 142:175-191.
5. Klas C, Debatin KM, Jonker RR et al. Activation interferes with the APO-1 pathway in mature human T cells. Int Immunol 1993; 5:625-630.
6. Lenardo MJ. Interleukin-2 programs mouse alpha beta T lymphocytes for apoptosis. Nature 1991; 353:858-861.
7. Watanabe-Fukunaga R, Brannan CI, Copeland NG et al. Lymphoproliferation disorder in mice explained by defects in Fas antigen that mediates apoptosis. Nature 1992; 356:314-317.
8. Takahashi T, Tanaka M, Brannan CI et al. Generalized lymphoproliferative disease in mice, caused by a point mutation in the Fas ligand. Cell 1994; 76:969-976.
9. Renno T, Hahne M, Tschopp J et al. Peripheral T cells undergoing superantigen-induced apoptosis in vivo express B220 and upregulate Fas and Fas ligand. J Exp Med 1996; 183:431-437.
10. Adachi M, Suematsu S, Kondo T et al. Targeted mutation in the Fas gene causes hyperplasia in peripheral lymphoid organs and liver. Nat Genet 1995; 11:294-300.
11. Rieux-Laucat F, Le Deist F, Fischer A. Autoimmune lymphoproliferative syndromes: genetic defects of apoptosis pathways. Cell Death Differ 2003; 10:124-133.
12. Werlen G, Hausmann B, Naeher D et al. Signaling life and death in the thymus: timing is everything. Science 2003; 299:1859-1863.
13. Surh CD, Sprent J. T-cell apoptosis detected in situ during positive and negative selection in the thymus. Nature 1994; 372:100-103.

14. Herron LR, Eisenberg RA, Roper E et al. Selection of the T cell receptor repertoire in Lpr mice. J Immunol 1993; 151:3450-3459.

15. Singer GG, Abbas AK. The fas antigen is involved in peripheral but not thymic deletion of T lymphocytes in T cell receptor transgenic mice. Immunity 1994; 1:365-371.

16. Geppert TD, Wacholtz MC, Davis LS et al. Activation of human T4 cells by cross-linking class I MHC molecules. J Immunol 1988; 140:2155-2164.

17. Alderson MR, Armitage RJ, Maraskovsky E et al. Fas transduces activation signals in normal human T lymphocytes. J Exp Med 1993; 178:2231-2235.

18. Kennedy NJ, Kataoka T, Tschopp J et al. Caspase Activation Is Required for T Cell Proliferation. J Exp Med 1999; 190:1891-1896.

19. Newton K, Harris AW, Bath ML et al. A dominant interfering mutant of FADD/MORT1 enhances deletion of autoreactive thymocytes and inhibits proliferation of mature T lymphocytes. EMBO J 1998; 17:706-718.

20. Zhang J, Cado D, Chen A et al. Fas-mediated apoptosis and activation-induced T-cell proliferation are defective in mice lacking FADD/Mort1. Nature 1998; 392:296-300.

21. Walsh CM, Wen BG, Chinnaiyan AM et al. A role for FADD in T cell activation and development. Immunity 1998; 8:439-449.

22. Zornig M, Hueber AO, Evan G. P53-dependent impairment of T-cell proliferation in FADD dominant-negative transgenic mice. Curr Biol 1998; 8:467-470.

23. Lens SM, Kataoka T, Fortner KA et al. The caspase 8 inhibitor c-FLIP(L) modulates T-cell receptor-induced proliferation but not activation-induced cell death of lymphocytes. Mol Cell Biol 2002; 22:5419-5433.

24. Salmena L, Lemmers B, Hakem A et al. Essential role for caspase 8 in T-cell homeostasis and T-cell-mediated immunity. Genes Dev 2003; 17:883-895.

25. Newton K, Strasser A. Caspases signal not only apoptosis but also antigen-induced activation in cells of the immune system. Genes Dev 2003; 17:819-825.

26. Altman A, Theofilopoulos AN, Weiner R et al. Analysis of T cell function in autoimmune murine strains. Defects in production and responsiveness to interleukin 2. J Exp Med 1981; 154:791-808.

27. Landolfi MM, Van Houten N, Russell JQ et al. CD2-CD4-CD8- lymph node T lymphocytes in MRL lpr/lpr mice are derived from a CD2+CD4+CD8+ thymic precursor. J Immunol 1993; 151:1086-1096.

28. Green DR, Droin N, Pinkoski M. Activation-induced cell death in T cells. Immunol Rev 2003; 193:70-81.

29. Hildeman DA, Zhu Y, Mitchell TC et al. Molecular mechanisms of activated T cell death in vivo. Curr Opin Immunol 2002; 14:354-359.

30. Mercep M, Weissman AM, Frank SJ et al. Activation-driven programmed cell death and T cell receptor zeta eta expression. Science 1989; 246:1162-1165.

31. Shi YF, Sahai BM, Green DR. Cyclosporin A inhibits activation-induced cell death in T-cell hybridomas and thymocytes. Nature 1989; 339:625-626.

32. Dhein J, Walczak H, Baumler C et al. Autocrine T-cell suicide mediated by APO-1/(Fas/CD95). Nature 1995; 373:438-441.

33. Brunner T, Mogil RJ, LaFace D et al. Cell-autonomous Fas (CD95)/Fas-ligand interaction mediates activation-induced apoptosis in T-cell hybridomas. Nature 1995; 373:441-444.

34. Ju ST, Panka DJ, Cui H et al. Fas(CD95)/FasL interactions required for programmed cell death after T-cell activation. Nature 1995; 373:444-448.

35. Alderson MR, Tough TW, Davis-Smith T et al. Fas ligand mediates activation-induced cell death in human T lymphocytes. J Exp Med 1995; 181:71-77.

36. Yang Y, Mercep M, Ware CF et al. Fas and activation-induced Fas ligand mediate apoptosis of T cell hybridomas: inhibition of Fas ligand expression by retinoic acid and glucocorticoids. J Exp Med 1995; 181:1673-1682.

37. Kawabe Y, Ochi A. Programmed cell death and extrathymic reduction of Vbeta8+ CD4+ T cells in mice tolerant to Staphylococcus aureus enterotoxin B. Nature 1991; 349:245-248.

38. Vella AT, McCormack JE, Linsley PS et al. Lipopolysaccharide interferes with the induction of peripheral T cell death. Immunity 1995; 2:261-270.

39. Mogil RJ, Radvanyi L, Gonzalez-Quintial R et al. Fas (CD95) participates in peripheral T cell deletion and associated apoptosis in vivo. Int Immunol 1995; 7:1451-1458.

40. Renno T, Hahne M, Tschopp J et al. Peripheral T cells undergoing superantigen-induced apoptosis in vivo express B220 and upregulate Fas and Fas ligand. J Exp Med 1996; 183:431-437.

41. Miethke T, Vabulas R, Bittlingmaier R et al. Mechanisms of peripheral T cell deletion: anergized T cells are Fas resistant but undergo proliferation-associated apoptosis. Eur J Immunol 1996; 26:1459-1467.

42. Zheng L, Fisher G, Miller RE et al. Induction of apoptosis in mature T cells by tumour necrosis factor. Nature 1995; 377:348-351.

43. Hildeman DA, Zhu Y, Mitchell TC et al. Activated T cell death in vivo mediated by proapoptotic bcl-2 family member bim. Immunity 2002; 16:759-767.

44. Kim C, Siminovitch KA, Ochi A. Reduction of lupus nephritis in MRL/lpr mice by a bacterial superantigen treatment. J Exp Med 1991; 174:1431-1437.

45. Zimmermann C, Rawiel M, Blaser C et al. Homeostatic regulation of CD8+ T cells after antigen challenge in the absence of Fas (CD95). Eur J Immunol 1996; 26:2903-2910.

46. Bouillet P, Metcalf D, Huang DC et al. Proapoptotic Bcl-2 relative Bim required for certain apoptotic responses, leukocyte homeostasis, and to preclude autoimmunity. Science 1999; 286:1735-1738.

47. Strasser A, Harris AH, Huang DCS et al. Bcl-2 and Fas/APO-1 regulate distinct pathways to lymphocyte apoptosis. EMBO J 1996; 14:6136-6147.

48. Smith KGC, Strasser A, Vaux DL. CrmA expression in T lymphocytes of transgenic mice inhibits CD95 (fas/APO-1)-transduced apoptosis, but does not cause lymphadenopathy or autoimmune disease. EMBO J 1996; 15:5167-5176.

49. Rescigno M, Piguet V, Valzasina B et al. Fas engagement induces the maturation of dendritic cells (DCs), the release of interleukin (IL)-1beta, and the production of interferon gamma in the absence of IL-12 during DC-T cell cognate interaction: A new role for Fas ligand in inflammatory responses. J Exp Med 2000; 192:1661-1668.

50. Restifo NP. Not so Fas: Reevaluating the mechanisms of immune privilege and tumor escape. Nat Med 2000; 6:493-495.

51. Bellgrau D, Gold D, Selawry H et al. A role for CD95 ligand in preventing graft rejection. Nature 1995; 377:630-632.

52. Griffith TS, Brunner T, Fletcher SM et al. Fas ligand-induced apoptosis as a mechanism of immune privilege. Science 1995; 270:1189-1192.

53. Allison J, Georgiou HM, Strasser A et al. Transgenic expression of CD95 ligand on islet beta cells induces a granulocytic infiltration but does not confer immune privilege upon islet allografts. Proc Natl Acad Sci USA 1997; 94:3943-3947.

54. Vaux DL. Immunology. Ways around rejection. Nature 1998; 394:133.

55. Green DR, Ferguson TA. The role of Fas ligand in immune privilege. Nat Rev Mol Cell Biol 2001; 2:917-924.

56. Wahlsten JL, Gitchell HL, Chan CC et al. Fas and Fas ligand expressed on cells of the immune system, not on the target tissue, control induction of experimental autoimmune uveitis. J Immunol 2000; 165:5480-5486.

57. Restifo NP. Countering the 'counterattack' hypothesis. Nat Med 2001; 7:259-

58. Niederkorn JY. The immune privilege of corneal grafts. J Leukoc Biol 2003; 74:167-171.

59. Takeda K, Hayakawa Y, Smyth MJ et al. Involvement of tumor necrosis factor-related apoptosis-inducing ligand in surveillance of tumor metastasis by liver natural killer cells. Nat Med 2001; 7:94-100.

60. Lowin B, Hahne M, Mattmann C et al. Cytolytic T-cell cytotoxicity is mediated through perforin and Fas lytic pathways. Nature 1994; 370:650-652.

61. Schmaltz C, Alpdogan O, Horndasch KJ et al. Differential use of Fas ligand and perforin cytotoxic pathways by donor T cells in graft-versus-host disease and graft-versus-leukemia effect. Blood 2001; 97:2886-2895.

CHAPTER 10

The FasL-Fas System in Disease and Therapy

Harald Wajant and Frank Henkler

Abstract

The physiological roles of the FasL-Fas system include apoptosis-related processes like tumor surveillance, elimination of virus infected cells or deletion of autoreactive T- and B-cells. In recent years there is also emerging evidence for nonapoptotic functions of these molecules in liver regeneration, T-cell activation and neurite differentiation. Disregulation or malfunction of FasL or Fas have been implicated in a variety of pathological situations including autoimmune diseases, fulminant hepatitis, graft versus host disease and spinal cord injury. This chapter is focused on current concepts of either inhibition or activation of Fas signaling as a therapeutic mean. Especially strategies that avoid the severe side effects of systemic Fas activation are discussed.

Physiological Role of the FasL-Fas System

FasL and Fas in Deletion of Auto-Reactive T- and B-Cells, Activation Induced Cell Death and T-Cell Activation

The in vivo function of FasL and Fas have been mainly defined by analysis of the generalized lymphoproliferative disease (gld) and lymphoproliferation (lpr) mutations in mice which impair FasL or Fas function, respectively.[1] The pathology of lpr and gld mice suggests that T- and B-cells escape from tolerance induction to self antigen, suggesting a role for both FasL and Fas in apoptosis- and/or anergy induction in auto-reactive T- and B-cells. With respect to thymic selection there are contradictory data in the literature showing reduced but mostly normal selection in transgenic models of lpr mice (Table 1). While the role of FasL and Fas in thymic selection is rather unclear there is strong evidence for a crucial role of these molecules in peripheral deletion of T-cells. When T-cells are repeatedly stimulated via the TCR or challenged with superantigen, apoptosis can be triggered by the FasL-Fas system.[2-4] This process is termed activation induced cell death (AICD). Thus, in lpr and gld mice superantigen treatment failed to delete the corresponding T-cell subsets and deletion of peripheral T-cells was abrogated in TCR transgenic mice after injection of a TCR-"specific" peptide antigen.[5-8] Remarkably, there is evidence that FasL upregulation in liver and intestine, rather than directly in T-cells, was responsible for cell death induction of activated peripheral T-cells.[5,9] Accumulation of CD4⁻CD8⁻ double negative T-cells in lpr mice is strongly reduced after crosslinking with MHC I deficient mice or upon long-term treatment with anti-CD8 antibodies suggesting that these cells were derived from CD8⁺ T-cells.[10-12] Activation, tolerance induction and AICD are normal in CD8⁺ T-cells. Moreover, there is evidence that

Fas Signaling, edited by Harald Wajant. ©2006 Landes Bioscience and Springer Science+Business Media.

Table 1. Pros and cons for a role of the FasL-Fas system in thymic selection

Study	Pros	Cons
Castro et al, 1996[1]	Reduced T-cell receptor-mediated apoptosis in lpr mice, Inhibition of antigen-induced apoptosis of thymocytes in non-lpr mice by-Fas-Fc	
Trimble et al, 2002[2]	Defect elimination of nonselected CD8 cells in lpr mice	
Kishimoto et al, 1997[3]	Defect elimination of semi-mature medullary T-cells at high doses of antigen in lpr mice	
Kishimoto and Sprent, 2001[4]	TCR-CD28 co-stimulation protects NOD but not normal tymocytes against Fas-mediated apoptosis	
Adachi et al, 1996[5]		Normal tymic selection in Fas knock out mice
Sidman et al, 1992[6]		Normal tymic selection in gld mice mice
Kotzin et al, 1988[7]		Normal repertoire modification with respect to potentially self-reactive TCR specificities in lpr mice
Mountz et al, 1990[8]		Superantigen corresponding T-cell subsets are absent in superantigen exprressing lpr mice
Giese and Davidson, 1992[9]		No chances in autoimmune prone T-cell subsets in lpr and gld mice
Singer and Abbas, 1994[10]		Normal thymocyte deletion after antigenic stimulation in lpr mice expressing a pigeon cytochrome c-specific TCR
Sytwu et al, 1996[11]		Normal thymocyte deletion after antigenic stimulation in lpr mice expressing a (HA)-specific TCR

[1]Castro JE, Listman JA, Jacobson BA et al. Immunity 1996; 5:617-627. [2]Trimble LA, Prince KA, Pestano GA et al. J Immunol 2002; 168:4960-4967. [3]Kishimoto H, Surh CD, Sprent J. J Exp Med 1998; 187:1427-1438. [4]Kishimoto H, Sprent J. Nat Immunol 2001; 2:1025-1031. [5]Adachi M, Suematsu S, Suda T et al. PNAS USA 1996; 93:2131-2136. [6]Sidman CL, Marshall JD, Von Boehmer H. Eur J Immunol 1992; 22:499-504. [7]Kotzin BL, Babcock SK, Herron LR. J Exp Med 1988; 168:2221-2229. [8]Mountz JD, Smith TM, Toth KS. J Immunol 1990; 144:2159-2166. [9]Giese T, Davidson WF. J Immunol 1992; 149: 3097-3106. [10]Singer GG, Abbas AK. Immunity 1994; 1:365-371. [11]Sytwu HK, Liblau RS, McDevitt HO. Immunity 1996; 5:17-30.

the failure of CD8[+] T-cells to interact with MHC I molecules leads to loss of CD8 expression and subsequent deletion via the FasL-Fas system.[13,14] Autoantibody production in lpr and gld mice is dependent on T-helper cells and relies most likely on escape of chronically desensitized autoreactive B-cells from Fas-induced cell death whereas the induction of B-cell tolerance seem to be normal in lpr mice.[15] Taken together, the FasL-Fas system has a central role in holding activated or potential autoreactive mature T- and B-cells in check while its role in negative selection during development of T- and B-cells seems to be less relevant.

Already shortly after identification of Fas as a death inducing receptor, it has been demonstrated that activation of Fas also enhances T-cell activation.[16] Moreover, a FasL neutralizing Fas-Fc fusion protein interfered with T-cell activation by suboptimal amounts of anti-CD3 indicating that TCR-induced endogenous FasL can contribute to T-cell activation.[17] Noteworthy, neutralization of the FasL-Fas interaction has no effect on T-cell activation induced by high amounts of anti-CD3. As lpr and gld mice display no major defects in T-cell activation the role of the FasL-Fas system in this process has to be nonessential and appears of only supportive nature. However, loss of expression of the Fas-associated proteins FADD and caspase-8 or pharmacological caspase inhibitors interfere with T-cell activation in vitro and in vivo.[17-23] Especially, in these cases T-cell activation by high doses of anti-CD3 is also blocked. Thus, it appears possible that Fas redundantly act with other death receptors in T-cell activation or it might only contribute to a yet nondefined TCR-induced FADD- and caspase-dependent, but death receptor-independent proliferative pathway.

FasL and Fas in Immune Privilege, Tumor Counterattack and Tumor Surveillance

In some tissue and organs in the body immune reactions are suppressed, a phenomenon called immune privilege.[24] The prototypic and best studied immune privileged organ is the eye. The immune suppressed status of the eye is for example reflected by the high acceptance of corneal allografts in the absence of immuno-suppressive therapy. Other immune privileged sites are brain, testis, ovary, placenta and the pregnant uterus. The mechanisms underlying immune privilege are complex and include physical barriers and local production of immune suppressive cytokines e.g., TGFβ.[24] In particular, FasL expression has been proposed as a mechanism to confer immune privilege by inducing apoptosis in Fas expressing inflammatory cells entering an immune privileged site. In accordance a variety of studies argue for a special role of FasL expression in the cornea for preventing inflammation and rejection of transplanted corneas. Firstly, Fas sensitive cells undergo apoptosis on cultured corneas via the Fas pathway.[25,26] Secondly, wt but not lpr T-cells undergo apoptosis upon injection into eyes of wt mice but not in eyes of gld mice.[27] Thirdly, inflammation in the eye induced by herpes simplex virus or toxoplasma is more severe in gld mice than in wt mice.[25,28] Fourthly, acceptance of grafted wt corneas is significantly higher than acceptance of corneas from gld mice.[29]

In line with the concept of immune privilege, it has been proposed that tumors utilize in a similar fashion FasL expression to escape immune responses. In fact, some Fas-resistant tumors constitutively express FasL and kill Fas sensitive cells in vitro.[30-33] Moreover, apoptosis of tumor infiltrating lymphocytes has been observed in situ in FasL expressing tumors.[32,34,35] In accordance with the FasL-mediated inhibition of an anti-tumoral immune response, enhanced growth of FasL positive tumors has also been observed, except in lpr mice, where a reduced growth of a FasL positive murine melanoma cell line had been reported.[32,36,37] However, a number of studies came to contradictory conclusions, describing accelerated rejection of FasL expressing grafts and rapid destruction of FasL positive tumors.[38-42] Obviously, secondary yet poorly defined factors are likely to regulate in vivo the effects of FasL expression on tumors, causing this controversial situation. FasL-mediated rejection occurs in human cells resistant to Fas-mediated apoptosis and seems to be dependent in the main on infiltrating neutrophils.[38-42] In agreement with a crucial role of neutrophils in FasL-induced tumor suppression, rejection of FasL-positive tumor cells was diminished in mice deficient for the neutrophil chemoattractant MIPα1.[43] After initial rejection of FasL expressing tumor cells, long time immunity against the tumor has also been observed in some studies. Tumor immunity can be based on a CD8+ response.[39,41] Further, tumor antigen specific antibodies can also induce complement mediated lysis of tumor cells in concert with CD4[+] T-cell help.[43] It is clear that neutrophils can kill tumor cells but the initial mechanisms leading to FasL-mediated neutrophil recruitment are complex and poorly understood. Early in vitro studies had suggested that soluble FasL acts directly as a chemotactic factor for neutrophils.[44,45] However, later studies failed to confirm a chemotactic activity of soluble FasL or found that expression of a noncleavable deletion mutant of membrane FasL is sufficient to induce neutrophil recruitment.[42,46-48] Noteworthy, soluble FasL expressing tumor cells failed to elicit a neutrophilic response in these studies. Thus, FasL-induced recruitment of neutrophiles in vivo occurs independent from a direct chemotactic action of FasL and has to depend on other mechanisms. It has been therefore proposed that FasL expressed on tumor cells triggers indirectly inflammation by induction of chemoattractant proteins in the surrounding cells. In this regard, it has been shown that FasL can induce IL8 and other chemokines in peritoneal macrophages, fibroblasts and various cell lines (see Chapter 6). Especially, FasL may induce IL1, which in turn can stimulate the production of neutrophil attractive factors.[49] In addition, FasL may boost an immune response by inducing dendritic cell (DC) maturation and by facilitating the uptake of tumor antigens by DCs.[50-52] The signaling pathways mediating the proinflammatory effects of FasL are complex and discussed in details in chapter 6.

Excessive proliferation does not necessarily leads to tumor formation as growth promoting signals also sensitize cells of the immune system to induce apoptosis in rapidly growing cells—a regulatory principle called tumor surveillance.[53] Thus, a tumor only develops when it can circumvent the tumor surveillance-related constrains, e.g., by acquisition of anti-apoptotic molecules or immuno-suppressive barriers. The inhibitory action of the immune system on tumorigenesis also delivers a rationale for the development of immunotherapy approaches in cancer treatment. Tumor surveillance is mainly based on the action of T-cells and NK cells. Both type of cells use two independent systems to induce apoptosis in cancer-prone cells. Firstly, calcium-dependent exocytosis of granule filled with perforin and granzymes, which cooperatively trigger apoptosis in target cells.[53] Secondly, apoptosis-induction by FasL and the related death ligand TRAIL.[54,55] Remarkably, the tumor suppressive effect of FasL and TRAIL may not only rely on direct induction of apoptosis in sensitive cells but may also be indirectly mediated by inducing proinflammatory cytokines and chemokines which establish a secondary local immune response. As already discussed above, such tumor suppressive inflammatory response might not only be triggered by FasL expressed on T-cells and NK cells, but also by expression of FasL by the tumor cells itself. Thus, the proinflammatory effects of FasL, which can occur independently from apoptosis, could possibly restrict the effects of FasL in the tumor counterattack. In accordance with a role of the FasL-Fas system in tumor surveillance, older lpr and gld mice develop plasmacytoid tumors and cross-breading of lpr or gld mice with mice that were predisposed to tumors showed a significantly increased tumor incidence.[56-59] Moreover, melanoma metastasis was found to be suppressed in gld, as compared to wt mice.[60,61] The anti-tumoral action of the FasL-Fas system is also in agreement with the finding that the FasL-Fas system is often impaired in tumor cells by mutations or inhibition of Fas transcription. The anti-tumoral effects of IL12 observed in some murine models have also been partly ascribed to Fas.[62,63] However, there is evidence that TRAIL is more relevant than FasL in these scenarios.[54,55]

FasL and Fas in Nerve Regeneration and Neurite Differentiation

A variety of cells of the nerveous system and the brain express Fas and / or FasL. According to the proapoptotic properties of the FasL-Fas system these molecules have been implicated in the pathology of various injuries of the nervous system including spinal cord injury and autoimmune diseases, like EAE.[64] Recently, it has been found that Fas signals ERK activation and neurite outgrowth in vitro and enhances peripheral nerve regeneration in vivo.[65,66] These findings shed new light on former studies reporting in lpr mice strial dysfunction, development of cochlear disease, cognitive deficits, delayed regeneration of neuronal root ganglia and progressive atrophy of pyramidal neuron dendrites.[67] Originally, most of these effects have been regarded as indirect consequences of autoimmunity developing in lpr mice. However, now it appears possible that at least some of these defects reflect loss of survival and regenerative functions of the FasL-Fas system. In fact, some of the neurological defects in lpr mice manifest before onset of autoimmunity and immuno-suppressive therapy inhibit the development of autoimmunity, but has no effect on the neurological phenotype of lpr mice.[68-70] The beneficial effects of Fas signaling on nerve regeneration may also explain why T-cell infiltration after traumatic neuronal injury is neuroprotective.[71] Further studies will be necessary to distinguish between the neurotoxic and neuroprotective functions of the FasL-Fas system.

FasL and Fas in Angiogenesis

Fas-induced apoptosis has also been implicated in the regulation of angiogenesis. For several inhibitors of angiogenesis including 2-methoxyestradiol, angiostatin, endostatin, thrombospondin-1 (TSP1) and pigment epithelium-derived factor (PEDF) apoptosis induction in cultured endothelial cells has been described.[72-75] The inhibitory effect of these listed inhibitors of angiogenesis was found to be dependent on caspase activation and was associated with apoptosis induction.[75] In particular, neutralization of FasL or Fas and selective inhibitors of caspase-8 prevented TSP1- and PEDF-induced apoptosis of endothelial cells.[75] Angiogenesis inhibitor induced apoptosis was in this scenario accompanied by upregulation of FasL. Remarkably, in the presence of inducers of neovascularization (e.g., IL8, bFGF or VEGF), which enhance cell surface trafficking of Fas, TSP1- and PEDF-induced apoptosis was significantly increased.[75] Moreover, in a corneal neovascularization assay TSP1 and PEDF mediate inhibition of bFGF-induced neovascularization in vivo in wt but not in lpr mice. Similarily, Fas-Fc reverts TSP1-mediated neovascularization in wt mice.[75] Besides Fas, angiogenesis inducers can also upregulate FLIP which is a potent inhibitor of Fas-induced apoptosis. Vice versa inhibitors of angiogenesis can block FLIP expression by inhibition of the Akt pathway.[76-78] Thus, a model of FasL-Fas-regulated angiogenesis arises, in which the interplay of angiogenesis inhibitor driven FasL expression and angiogenesis inducer-mediated upregulation of Fas and FLIP determines the extent of angiogenesis induced by a mixture of pro- and anti-angiogenic proteins. Especially, the short half-life of FLIP (see Chapter 6) could convert a pro-angiogenic overall response into an anti-angiogenic state, when angiogenesis inducing molecules are vanished.

Fas Signaling Induces Cardiac Hypertrophy

Pressure overload, e.g., as a consequence of arterial hypertension, induces adaptive cardiac growth as a compensatory response. However, sustained cardiac hypertrophy can lead to heart failure. Fas expression is upregulated in the myocard after volume overload induced cardiac hypertrophy and in biopsies of failing human hearts.[79] Moreover, in wt but not lpr mice induction of pressure overload by abdominal aortic constriction elicits typical features of a cardic hypertrophic response including an increase in septal as well as posterior wall thickness and in the ratio of heart to body weight.[80] Consequently, lpr mice showed a more than 3 fold increased lethality rate after aortic banding.[80] In cultured neonatal cardiomyocytes FasL induces the characteristic molecular hallmarks of the hypertrophic response, as increased total protein synthesis or production of the atrial nariuretic factor to a similar extent as the well established hypertrophic factors angiotensin II and endothelkin-1.[80] In vitro the hypertrophic response of FasL relies on Akt-mediated inhibition of glycogen synthase kinase 3β and subsequent NFAT-mediated transcription of hypertrophy-related genes. Fas-induced activation of JNK may have an additional role in the hypertrophic response as this pathway has also been implicated in the latter and JNK activation occurs after aortic binding in wt, but not lpr mice.[80-82]

Pathophysiology of FasL and Fas

Autoimmue Diseases

The pathophysiological consequences of defects in Fas or FasL are evident from lpr and gld mutant mice that express loss of function mutants of Fas and FasL. Both mutations are associated with autoimmunity.[1] The lpr phenotype is based on a retroviral insertion into the Fas gene leading to premature termination of transcription (lpr) or on a point mutation in the death domain of Fas (lpr-cg).[83] The gld phenotype is caused by a point mutation in the FasL gene. FasL- and Fas-deficient mice have in some aspects a more severe phenotype than lpr and gld mice suggesting that there is some residual FasL- and Fas function in these mutant mice.[84] A human counterpart to the lpr/gld phenotype in mice is known as autoimmune lymphoproliferative syndrome (ALPS).[85] In most ALPS patients, defects in the genes for Fas (ALPS Ia), FasL (APLS Ib) or in the Fas-associated protein caspase-10 have been found. Lpr and gld mice are both characterized by lymphoid hyperplasia, accumulation of CD4⁻ CD8⁻ T-cells, production of anti-DNA autoantibodies and hypergammaglobulinemia.[1,83,84] Dependent on the background the mutant mice can also develop a more severe lupus-like systemic autoimmune disease with glomerulonephritis, vasculitis and arthritis.[1,83,84] As elimination of activated peripheral T-cells is impaired in lpr and gld mice, it is no surprise that these animals develop autoimmune diseases. In particular, the FasL-Fas system has been implicated in the pathogenesis of autoimmune thyroiditis, multiple sclerosis, type 1 diabetes mellitus and experimental allergic encephalomyelitis (EAE).[15,86] A significant contribution of the FasL-Fas system to the pathogenesis of insulin-dependent diabetes mellitus is evident from nonobese diabetic lpr mice which in contrast to NOD mice do not develop hyperglycaemia and diabetes.[15,86] A more complex role of FasL has been recognized for the FasL-Fas system in experimental allergic encephalomyelitis (EAE), a model for multiple sclerosis—a demylinating autoimmune disease. On the one side, EAE has found to be reduced in lpr and gld mice arguing for a crucial role of the FasL-Fas system in the pathogeneses of this disease presumably by apoptosis induction in oligodendrocytes, leading subsequently to demyelination.[87-93] On the other side spontanous EAE was increased in mice with defects in the Fas or FasL, gene suggesting that spontaneous recovery from EAE is due to AICD of immune cells.[90,94]

FasL and Fas in Hepatits

Deregulated Fas signaling in the liver may induce hepatocyte apoptosis and has been implicated in hepatitis.[95,96] Chronic infection with hepatitis B virus (HBV) or hepatitis C virus (HCV), is associated with an enhanced Fas-mediated hepatocyte apoptosis. The immunopathological effects in viral hepatitis are mediated by CD8⁺ T-cells.[97] For this pathology FasL expression in liver-infiltrating CTLs seems to be of special relevance. because in a hepatitis B surface antigen transgenic mouse model liver damage was blocked by a FasL neutralizing Fas-Fc fusion protein.[98] Further, liver inflammation in viral hepatitis leads to an upregulation of Fas, thus to sensitization of hepatocytes for apoptosis induction. In chronic HCV infection, there is an apparent correlation between up-regulation of Fas and severity of inflammation.[99] The same applies to chronic HBV infection.[100] However, there is also evidence that FasL expression can be induced by viral proteins, especially by the hepatitis B X-protein which is suggested to play a role in viral replication.[101] Perhaps, FasL expression might selectively protect hepatocytes with ongoing HBV replication from CTL-mediated apoptosis. This mechanism could be further relevant in the progression of chronic HBV infection into carcinogenesis. Infact, FasL expression is detectable in a high proportion of hepatocellular carcinomas, whereas this protein is not expressed in normal liver.[102]

Blockage of FasL-Fas interaction also demonstrated a crucial role of these molecules in liver damage by Propionibacterium acnes and LPS.[98,103] Further, studies with gld and lpr mice implicated the FasL-Fas system in ConA-induced liver failure.[104,105] However, in this case the relevant FasL expressing cell seems to be NK T-cells. FasL and Fas may also have nonpathological functions in the liver. So, there is evidence from Fas ko mice that Fas-induced apoptosis contributes to the elimination of senescent hepatocytes.[106] Moreover, nonapoptotic Fas signaling enhances the regenerative proliferation of hepatocytes.[107]

The FasL-Fas System in Spinal Cord Injury

Fas and FasL are upregulated in a variety of traumatic and ischemic neurological disorders and often associated with increased apoptosis.[64] This may reflect a causal role for the FasL-Fas system in these injuries but could also be a secondary effect of an initially regenerative Fas response as described in detail in I.3. The explicit role of FasL and Fas in the various diseases remains therefore to be elucidated. With respect to motoneurons, the FasL-Fas system fulfils primarily proapoptotic functions.[108,109] After deprivation of trophic factors like BDNFs, a significant protion of cultured motoneurons undergoes Fas-mediated apoptosis.[100] Lpr and gld mice and trangenic mice overexpressing a dominant negative deletion mutant of FADD have no defects in programmed cell death of spinal motoneurons during development, although cultured motoneurons of these mice exert delayed apoptosis after trophic deprivation.[108,109] However, the principle Fas sensitivity of motoneurons observed in vitro becomes also apparent in vivo when axotomy-induced apoptosis of facial and spinal motoneurons were analyzed. In these scenarios mice with deficient Fas signaling showed improvements, but not a total rescue.[109] This also applies for the transgenic mice expressing dominant-negative FADD, suggesting that death receptor independent cell death plays a role in axotomy-induced apoptosis of motoneurons, too. The source of FasL acting in axotomy-induced apoptosis could either be motoneurons, or invading T-cells. Motoneurons derived from SOD1 mice, which are a model for amytrophic lateral sclerosis (ALS), showed increased Fas sensitivity opening the possibility that Fas-induced apoptosis is involved in the pathogenesis of neurodegenerative diseases, too.[110] The Fas signaling pathway, especially FasL and Fas, is therefore a possible point of intervention to prevent motoneuron death. In fact, neutralization of FasL by antibody treatment has shown to reduce the number of dead cells and enhance recovery after spinal cord injury.[111]

Therapeutic Potential of FasL and Fas Neutralizing Reagents and Fas Agonists

Neutralization of Fas

In diseases where Fas signaling critically contributes to the pathophysiological situation, neutralization of the FasL-Fas interaction could be an useful therapeutic option (Fig. 1). Long-term blocking of Fas signaling could lead to autoimmune complications but acute or transient blockage should be a possible way, as neutralizing reagents FasL-specific antibodies or soluble Fas variants could be used. With respect to the latter there is evidence that Fas-Comp, a pentameric fusion protein of Fas, has a significant higher neutralization capacity as the classical dimeric Fc fusion proteins.[112] A recent report described the generation of an exocyclic Fas mimicing peptide, which can block Fas-induced apoptosis in vitro and Fas-driven concavalin A-induced hepatitis in vivo.[113] It appears therefore possible that a therapeutically effective inhibition of the FasL-Fas interaction can also be achieved with low molecular weight compounds. In several mouse models of Fas-related pathologies, including models for spinal cord injury, EAE and hepatitis, the principle usefulness of FasL neutralization has been demonstrated (see

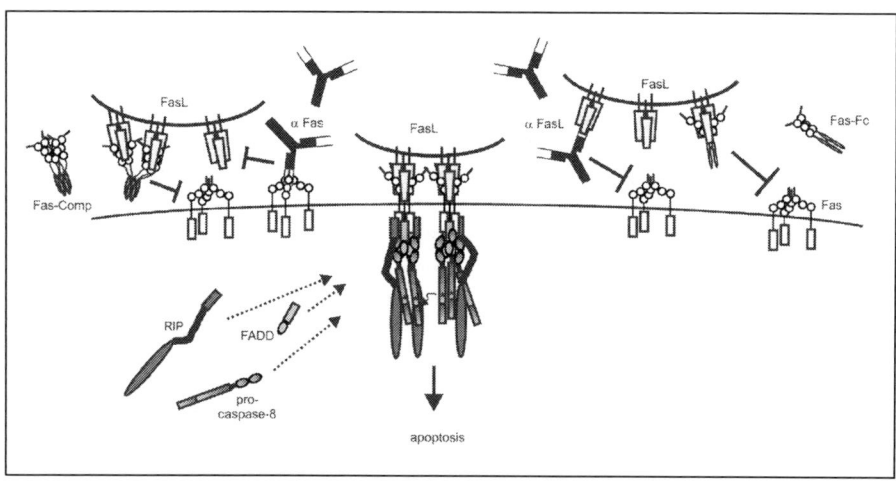

Figure 1. Inhibition of FasL-Fas interaction by antibodies recognizing the extracellular domain of FasL or Fas and soluble chimeric proteins containing the extracellular domain of Fas.

previous paragraphs). It remains to be seen whether FasL blocker will have a similar success as the inhibitors of the FasL related cytokine TNF.

Agonistic Fas-Specific Antibodies

As systemic activation of Fas by cross-linked FasL does induce fulminant fatal liver damage, agonistic Fas-specific mAbs intuitively can have no reasonable chance for a broad relevance in therapeutic applications. However, there is evidence that agonistic Fas-sepcific antibodies display unexpected selectivities in vivo. For example, intraperitoneal administration of the hamster anti-mouse Fas mAb Jo2 induces rapid liver failure and hemorrhage.[114] In contrast the hamster anti-mouse Fas mAb RK8 has no such deleterious effects in the liver, but nevertheless efficiently induces thymocyte apoptosis and thymic atrophy.[115,116] Moreover, RK8 treatment inhibits lymphoadenopathy and splenomegaly in gld mice. These qualitative different bioactivities of Jo2 and RK8 become also evident in vitro. While Jo2 kills cultured primary hepatocytes and thymocytes, RK8 can only trigger apoptosis efficiently in the latter.[115] Depending on whether the intrinsic apoptotic pathway significantly contributes to Fas-induced apoptosis or not, Fas-sensitive cells have been classified into type I and type II cells.[117] In type I cells, Fas robustly triggers the caspase-8 - caspase-3 cascade leading to rapid and direct execution of the apoptotic program, so that expression of Bcl2 or other inhibitors of the mitochondrial apoptotic branch have no effect. In contrast, in type II cells the release of proapoptotic proteins from mitochondria is necessary and does further enhance of the action of the caspase-8 caspase-3 cascade. In these cells regulators of the intrinsic apoptotic pathway can therefore modulate Fas-induced apoptosis. Hepatocytes have been suggested to be type II cells whereas thymocytes belong to the type I category. Thus, it seems possible that for unknown reasons the RK8 mAb predominantly act on type I cells. It is obvious that a human Fas directed mAb with similar properties as RK8 would be of potential therapeutic interest, as it might facilitate selective killing of autoreactive immune cells in autoimmune diseases without inducing liver toxicity. In this regard, the group of S. Yonehara has derived a RK8-like mAb recognizing Fas from various species including men and mice by immunization of Fas deficient mice with human Fas.[118] This antibody called HEF7A, shows no hepatotoxicity in mice, marmosets and crab-eating monkey but decreased the autoimmune phenotype of gld mice. Similar to RK8, this mAb seems to act predominantly on type I cells.[118]

FasL in Gene Therapy

As already discussed above, attempts to use transgenic FasL expression led to complex and, at the first glance, contradictory results. While FasL expression conferred in certain transplants and some tumors an "immune privileged" status, other tumors and transplants triggered inflammation and tissue destruction. Thus, the overall consequences of FasL gene transfer might be determined by opposing mechanisms and are therefore poorly predictable. This is particular evident from studies where FasL gene transfer has been analyzed as a potential mean of blocking autoimmune processes—an obvious idea regarding the autoimmune lymphoproliferative defects of Fas and FasL deficient mice. While FasL transgenic myoblasts improved the acceptance of islet allografts in one study, FasL expressing islets boosted transplant rejection.[119-123] Importantly, the major requirement for FasL based gene therapy concepts in treatment of autoimmue diseases is, besides circumvention of Fas-mediated inflammation, the prevention of Fas-induced systemic toxicity. The latter may not represent a major problem in therapy concepts for rheumtoid arthritis as in arthritic joints the inflammatory cells and tissue are "encapsulated" in a cavity. Infact, adenovirus mediated FasL-gene tranfer into inflamed joints triggered strong FasL expression, induced synovial cell apoptosis and reduced disease without systemic toxicity in a collagen-induced arthritis model.[124-126] Replication defective adenovirus also offers the possibility to overcome systemic FasL toxicity by the use of inducible tissue-specific promoters. While intravenous infection of adenovirus with a CMV promoter driven FasL gene led to the typical liver damage of systemic Fas activation, the use of neuronal-, astrocyte- or smooth muscle cell-specific promoters to regulate FasL expression resulted in nontoxic adenoviral vectors.[127-129] In vitro the usefulness of an adenoviral expression system containing the FasL gene under the control of a tetracyclin-regulated promoter has also been demonstrated, but its value has not yet been shown in vivo.[130-132]

Cell Surface Immobilization Dependent Activation of FasL Fusion Proteins

The tumor inhibitory effect of chemotherapy is in part dependent on p53-induced up-regulation of Fas and FasL.[133-135] As many tumors have defects in the p53 pathway, direct stimulation of Fas should be a powerful mean in anti-cancer therapy. However, all Fas activating reagents, including agonistic antibodies, multimeric recombinant FasL preparations and vesicles containing membrane FasL induce to some degree liver failure. Nevertheless, the principle usefulness of Fas activation as an anti-tumoral tool was evident from studies where systemic action of Fas was prevented. For example, in mice lacking a functional FasL-Fas system (gld/lpr mice) transplanted tumors with functional Fas were destroyed by agonistic antibodies and human Fas-specific antibodies suppressed growth of human xenotransplants in nude mice models.[136,137] Thus, tumor therapeutic concepts employing Fas activation have to be strictly limited to localized action and should be ideally only be effective at the tumor site. The finding that trimeric soluble FasL (sFasL) fails more or less to activate Fas, but only becomes potently bioactive after secondary crosslinking, has suggested that the sFasL-Fas interaction alone is not sufficient to drive formation of signaling competent Fas clusters. A secondary higher aggregation of FasL-trimers is apparently necessary for a full activation of Fas.[138] It has been suggested that cell surface immobilization of an inactive sFasL fusion protein by a Fas/FasL independent protein-protein interaction domain may convert such a recombinant soluble FasL fusion protein into a membrane FasL-analog reagent (Fig. 2). If such a protein-protein interaction-domain would selectively target a tumor marker, this would restrict the apoptotic effects of FasL to the tumor cells or to cells adjacent to the targeted tumor cells. The proof of principle for this concept has been demonstrated with a fusion protein containing the extracellular domain of FasL fused to the C-terminus of a single-chain antibody which binds the fibroblast activation protein (FAP).[139] The latter is a membrane protein abundantly expressed on activated fibroblasts that surround a variety of solid

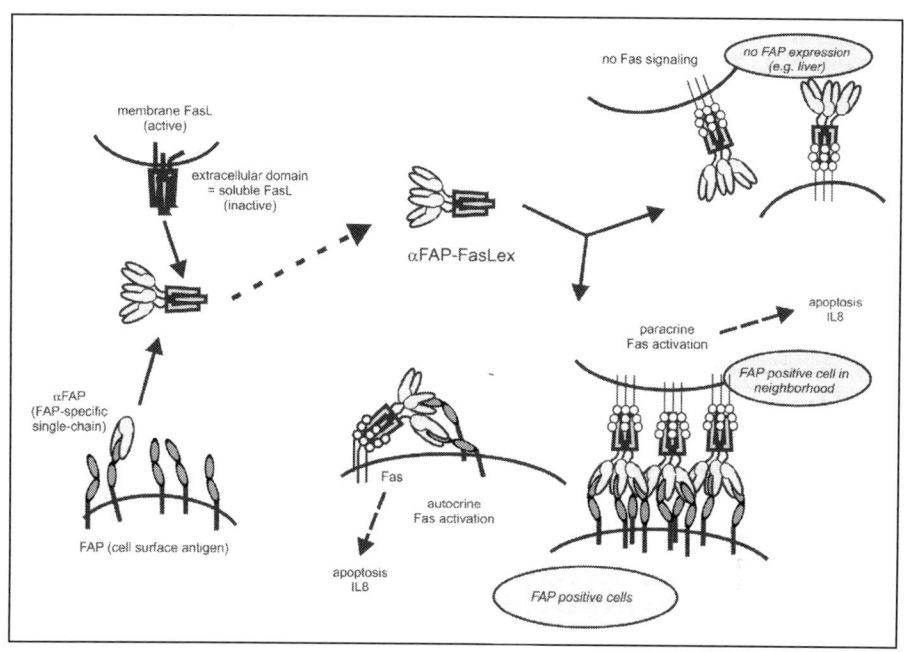

Figure 2. Cell surface immobilization dependent conversion of soluble trimeric FasL fusion proteins to entities with membrane FasL-like activity. A fusion protein of the extracellular domain of FasL and a single chain fragment recognizing the cell surface antigen fibroblast activation protein (aFAP-FasL) is unable to activate Fas unless it is bind to a cell surface via interaction with FAP.

tumors. In respect to apoptosis induction, this anti-FAP FasL fusion protein was on FAP expressing cells comparable or even more active as crosslinked sFasL, but showed a 1000 fold reduced activity on FAP negative cells. The FAP-dependent activation of Fas by this protein was evident from the inhibitory effect of a neutralizing FAP specific antibody.[139] In agreement with the FAP-dependent mode of Fas activation which was demonstrated in vitro, the anti-FAP-FasL fusion protein showed no signs of toxicity in mice, even after injection of 100 µg per mouse, but it blocked selectively the development of FAP positive xenotransplants.[139] Enhanced bioactivities of similar single chain fusion proteins containing the FasL-related cytokines TNF and TRAIL have also been demonstrated.[140,141] The superior functional capabilities of the TNF- and TRAIL fusion proteins, as compared to soluble TNF or soluble TRAIL, are based on activation of TNF-R2 and TRAIL-R2. Similar to Fas these receptors were predominantly activated by the membrane bound forms of their corresponding ligands. As soluble TNF and soluble TRAIL are already able to fully activate TNF-R1 and TRAIL-R1, respectively, a cell surface immobilization mediated increase in activity is highly dependent on the receptor expression pattern in a given cell. Taken together, this novel approach of artificial cell surface immobilization of soluble fusion proteins, containing a receptor-binding domain of a ligand of the TNF ligand family that is fused to an unrelated cell surface protein-binding domain, can be used to increase the signalling capabilities of various TNF ligands. As this immobilisation depends on the association between the cell surface protein binding domain and its tumor- or cell type specific target, a local activation of the corresponding receptors can be achieved, and systemic side effects should be avoided. Cell surface immobilization may also convert bispecific antibodies containing a nonagonistic Fas-specific antibody fragment into Fas activating entities.[142]

Outlook: Avoidance of Side Effects of Systemic Fas Activation by Isolate Limb Perfusion or Prodrug Strategies

Induction of severe side effects upon systemic application is not only a problem with Fas agonists, but also limits the therapeutic potential of other cytokines for example TNF. Thus, strategies that have been developed in recent years to achieve locale TNF activity, could possibly also used for FasL. Two such strategies are the isolated limb perfusion and the generation of inactive ligand fusion proteins that become activated by tumor associated proteases. The isolated limb perfusion has been developed to treat soft tissue sarcoma of the limbs with high doses of TNF and mephalane.[143,144] To prevent TNF from inducing shock-like complications in the body, the respective limb was disconnected from the circulatory system and externally supplied with oxygen during treatment. It is obvious that the isolated limb perfusion could also be used with Fas agonists. While the ILP is already in clinical use, the TNF prodrug strategy is in an early experimental state. The TNF prodrug concept is based on the generation of an intact TNF fusion protein in which TNF action is blocked by a c-terminal inhibitory domain consisting of the extracellular ligand-binding domain of TNF-R1.[141] As the TNF and TNFR1 domains are connected by a linker harbouring cleavage sites for proteases which are predominantly active in tumors, the inhibitory domain can be removed in the microenvironment of a tumor. To facilitate tumor associated activation and to prevent release of the processed active TNF from the tumor, the TNF-TNFR1 fusion protein contains in addition an N-terminal single chain domain, recognizing a tumor associated cell surface antigen. Similar FasL prodrugs fusion proteins may develop for applications, where cell surface immobilization dependent activation of FasL is not possible due to aggregating immobilization domains or ubiquitous expression of the available target structure.

Acknowledgement

This work was supported by Deutsche Forschungsgemeinschaft (Grant Wa 1025/11-1 Sonderforschungsbereich 495 project A5) and Dr. Mildred Scheel Stiftung für Krebsforschung grant 10-1751.

References

1. Cohen PL, Eisenberg RA. Lpr and gld: Single gene models of systemic autoimmunity and lymphoproliferative disease. Annu Rev Immunol 1991; 9:243-269.
2. Dhein J, Walczak H, Baumler C et al. Autocrine T-cell suicide mediated by APO-1/(Fas/CD95). Nature 1995; 373:438-441.
3. Brunner T, Mogil RJ, LaFace D et al. Cell-autonomous Fas (CD95)/Fas-ligand interaction mediates activation-induced apoptosis in T-cell hybridomas. Nature 1995; 373:441-444.
4. Ju ST, Panka DJ, Cui H et al. Fas(CD95)/FasL interactions required for programmed cell death after T-cell activation. Nature 1995; 373:444-448.
5. Bonfoco E, Stuart PM, Brunner T et al. Inducible nonlymphoid expression of Fas ligand is responsible for superantigen-induced peripheral deletion of T cells. Immunity 1998; 9:711-720.
6. Mogil RJ, Radvanyi L, Gonzalez-Quintial R et al. Fas (CD95) participates in peripheral T cell deletion and associated apoptosis in vivo. Int Immunol 1995; 7:1451-1458.
7. Singer GG, Abbas AK. The fas antigen is involved in peripheral but not thymic deletion of T lymphocytes in T cell receptor transgenic mice. Immunity 1994; 1:365-371.
8. Sytwu HK, Liblau RS, McDevitt HO. The roles of Fas/APO-1 (CD95) and TNF in antigen-induced programmed cell death in T cell receptor transgenic mice. Immunity 1996; 5:17-30.
9. Pinkoski MJ, Droin NM, Lin T et al. Nonlymphoid Fas ligand in peptide-induced peripheral lymphocyte deletion. Proc Natl Acad Sci USA 2002; 99:16174-16179.
10. Mixter PF, Russell JQ, Durie FH et al. Decreased CD4-CD8- TCR-alpha beta + cells in lpr/lpr mice lacking beta 2-microglobulin. J Immunol 1995; 154:2063-2074.

11. Ohteki T, Iwamoto M, Izui S et al. Reduced development of CD4-8-B220+ T cells but normal autoantibody production in lpr/lpr mice lacking major histocompatibility complex class I molecules. Eur J Immunol 1995; 25:37-41.

12. Merino R, Fossati L, Iwamoto M et al. Effect of long-term anti-CD4 or anti-CD8 treatment on the development of lpr CD4- CD8- double negative T cells and of the autoimmune syndrome in MRL-lpr/lpr mice. J Autoimmun 1995; 8:33-45.

13. Pestano GA, Zhou Y, Trimble LA et al. Inactivation of misselected CD8 T cells by CD8 gene methylation and cell death. Science 1999; 284:1187-1191.

14. Trimble LA, Prince KA, Pestano GA et al. Fas-dependent elimination of nonselected CD8 cells and lpr disease. J Immunol 2002; 168:4960-4967.

15. Lee HO, Ferguson TA. Biology of FasL. Cytokine Growth Factor Rev 2003; 14:325-335.

16. Alderson MR, Armitage RJ, Maraskovsky E et al. Fas transduces activation signals in normal human T lymphocytes. J Exp Med 1993; 178:2231-2235.

17. Kennedy NJ, Kataoka T, Tschopp J et al. Caspase activation is required for T cell proliferation. J Exp Med 1999; 190:1891-1896.

18. Alam A, Cohen LY, Aouad S et al. Early activation of caspases during T lymphocyte stimulation results in selective substrate cleavage in nonapoptotic cells. J Exp Med 1999; 190:1879-1890.

19. Mack A, Hacker G. Inhibition of caspase or FADD function blocks proliferation but not MAP kinase-activation and interleukin-2-production during primary stimulation of T cells. Eur J Immunol 2002; 32:1986-1992.

20. Newton K, Harris AW, Bath ML et al. A dominant interfering mutant of FADD/MORT1 enhances deletion of autoreactive thymocytes and inhibits proliferation of mature T lymphocytes. Embo J 1998; 17:706-718.

21. Walsh CM, Wen BG, Chinnaiyan AM et al. A role for FADD in T cell activation and development. Immunity 1998; 8:439-449.

22. Zhang J, Cado D, Chen A et al. Fas-mediated apoptosis and activation-induced T-cell proliferation are defective in mice lacking FADD/Mort1. Nature 1998; 392:296-300.

23. Chun HJ, Zheng L, Ahmad M et al. Pleiotropic defects in lymphocyte activation caused by caspase-8 mutations lead to human immunodeficiency. Nature 2002; 419:395-399.

24. Green DR, Ferguson TA. The role of Fas ligand in immune privilege. Nat Rev Mol Cell Biol 2001; 2:917-924.

25. Griffith TS, Brunner T, Fletcher SM et al. Fas ligand-induced apoptosis as a mechanism of immune privilege. Science 1995; 270:1189-1192.

26. Stuart PM, Griffith TS, Usui N et al. CD95 ligand (FasL)-induced apoptosis is necessary for corneal allograft survival. J Clin Invest 1997; 99:396-402.

27. Griffith TS, Yu X, Herndon JM et al. CD95-induced apoptosis of lymphocytes in an immune privileged site induces immunological tolerance. Immunity 1996; 5:7-16.

28. Hu MS, Schwartzman JD, Yeaman GR et al. Fas-FasL interaction involved in pathogenesis of ocular toxoplasmosis in mice. Infect Immun 1999; 67:928-935.

29. Yamagami S, Kawashima H, Tsuru T et al. Role of Fas-Fas ligand interactions in the immunorejection of allogeneic mouse corneal transplants. Transplantation 1997; 64:1107-1111.

30. Villunger A, Egle A, Marschitz I et al. Constitutive expression of Fas (Apo-1/CD95) ligand on multiple myeloma cells: A potential mechanism of tumor-induced suppression of immune surveillance. Blood 1997; 90:12-20.

31. O'Connell J, O'Sullivan GC, Collins JK et al. The Fas counterattack: Fas-mediated T cell killing by colon cancer cells expressing Fas ligand. J Exp Med 1996; 184:1075-1082.

32. Hahne M, Rimoldi D, Schroter M et al. Melanoma cell expression of Fas(Apo-1/CD95) ligand: Implications for tumor immune escape. Science 1996; 274:1363-1366.

33. Niehans GA, Brunner T, Frizelle SP et al. Human lung carcinomas express Fas ligand. Cancer Res 1997; 57:1007-1012.

34. Strand S, Hofmann WJ, Hug H et al. Lymphocyte apoptosis induced by CD95 (APO-1/Fas) ligand-expressing tumor cells—a mechanism of immune evasion? Nat Med 1996; 2:1361-1366.

35. Bennett MW, O'Connell J, O'Sullivan GC et al. The Fas counterattack in vivo: Apoptotic depletion of tumor-infiltrating lymphocytes associated with Fas ligand expression by human esophageal carcinoma. J Immunol 1998; 160:5669-5675.

36. Arai H, Chan SY, Bishop DK et al. Inhibition of the alloantibody response by CD95 ligand. Nat Med 1997; 3:843-848.
37. Nishimatsu H, Takeuchi T, Ueki T et al. CD95 ligand expression enhances growth of murine renal cell carcinoma in vivo. Cancer Immunol Immunother 1999; 48:56-61.
38. Yagita H, Seino K, Kayagaki N et al. CD95L in graft rejection. Nature 1996; 379:682.
39. Seino K, Kayagaki N, Okumura K et al. Antitumor effect of locally produced CD95 ligand. Nat Med 1997; 3:165-170.
40. Okamoto S, Takamizawa S, Bishop W et al. Overexpression of Fas ligand does not confer immune privilege to a pancreatic beta tumor cell line (betaTC-3). J Surg Res 1999; 84:77-81.
41. Shimizu M, Fontana A, Takeda Y et al. Induction of antitumor immunity with Fas/APO-1 ligand (CD95L)-transfected neuroblastoma neuro-2a cells. J Immunol 1999; 162:7350-7357.
42. Behrens CK, Igney FH, Arnold B et al. CD95 ligand-expressing tumors are rejected in anti-tumor TCR transgenic perforin knockout mice. J Immunol 2001; 166:3240-3247.
43. Simon AK, Gallimore A, Jones E et al. Fas ligand breaks tolerance to self-antigens and induces tumor immunity mediated by antibodies. Cancer Cell 2002; 2:315-322.
44. Seino K, Iwabuchi K, Kayagaki N et al. Chemotactic activity of soluble Fas ligand against phagocytes. J Immunol 1998; 161:4484-4488.
45. Ottonello L, Tortolina G, Amelotti M et al. Soluble Fas ligand is chemotactic for human neutrophilic polymorphonuclear leukocytes. J Immunol 1999; 162:3601-3606.
46. Shudo K, Kinoshita K, Imamura R et al. The membrane-bound but not the soluble form of human Fas ligand is responsible for its inflammatory activity. Eur J Immunol 2001; 31:2504-2511.
47. Hohlbaum AM, Moe S, Marshak-Rothstein A. Opposing effects of transmembrane and soluble Fas ligand expression on inflammation and tumor cell survival. J Exp Med 2000; 191:1209-1220.
48. Kang SM, Braat D, Schneider DB et al. A noncleavable mutant of Fas ligand does not prevent neutrophilic destruction of islet transplants. Transplantation 2000; 69:1813-1817.
49. Miwa K, Asano M, Horai R et al. Caspase 1-independent IL-1beta release and inflammation induced by the apoptosis inducer Fas ligand. Nat Med 1998; 4:1287-1292.
50. Rescigno M, Piguet V, Valzasina B et al. Fas engagement induces the maturation of dendritic cells (DCs), the release of interleukin (IL)-1beta, and the production of interferon gamma in the absence of IL-12 during DC-T cell cognate interaction: A new role for Fas ligand in inflammatory responses. J Exp Med 2000; 192:1661-1668.
51. Tada Y, OW J, Takiguchi Y et al. Cutting edge: A novel role for Fas ligand in facilitating antigen acquisition by dendritic cells. J Immunol 2002; 169:2241-2245.
52. Guo Z, Zhang M, An H et al. Fas ligation induces IL-1beta-dependent maturation and IL-1beta-independent survival of dendritic cells: Different roles of ERK and NF-kappaB signaling pathways. Blood 2003; 102:4441-4447.
53. Igney FH, Krammer PH. Death and anti-death: Tumour resistance to apoptosis. Nat Rev Cancer 2002; 2:277-288.
54. Wajant H, Pfizenmaier K, Scheurich P. TNF-related apoptosis inducing ligand (TRAIL) and its receptors in tumor surveillance and cancer therapy. Apoptosis 2002; 7:449-459.
55. Wang S, El-Deiry WS. TRAIL and apoptosis induction by TNF-family death receptors. Oncogene 2003; 22:8628-8633.
56. Davidson WF, Giese T, Fredrickson TN. Spontaneous development of plasmacytoid tumors in mice with defective Fas-Fas ligand interactions. J Exp Med 1998; 187:1825-1838.
57. Zornig M, Grzeschiczek A, Kowalski MB et al. Loss of Fas/Apo-1 receptor accelerates lymphomagenesis in E mu L-MYC transgenic mice but not in animals infected with MoMuLV. Oncogene 1995; 10:2397-2401.
58. Peng SL, Robert ME, Hayday AC et al. A tumor-suppressor function for Fas (CD95) revealed in T cell-deficient mice. J Exp Med 1996; 184:1149-1154.
59. Traver D, Akashi K, Weissman IL et al. Mice defective in two apoptosis pathways in the myeloid lineage develop acute myeloblastic leukemia. Immunity 1998; 9:47-57.
60. Owen-Schaub L, Chan H, Cusack JC et al. Fas and Fas ligand interactions in malignant disease. Int J Oncol 2000; 17:5-12.
61. Owen-Schaub LB, van Golen KL, Hill LL et al. Fas and Fas ligand interactions suppress melanoma lung metastasis. J Exp Med 1998; 188:1717-1723.

62. Wigginton JM, Lee JK, Wiltrout TA et al. Synergistic engagement of an ineffective endogenous anti-tumor immune response and induction of IFN-gamma and Fas-ligand-dependent tumor eradication by combined administration of IL-18 and IL-2. J Immunol 2002; 169:4467-4474.
63. Wigginton JM, Gruys E, Geiselhart L et al. IFN-gamma and Fas/FasL are required for the antitumor and antiangiogenic effects of IL-12/pulse IL-2 therapy. J Clin Invest 2001; 108:51-62.
64. Choi C, Benveniste EN. Fas ligand/Fas system in the brain: Regulator of immune and apoptotic responses. Brain Res Rev 2004; 44:65-81.
65. Shinohara H, Yagita H, Ikawa Y et al. Fas drives cell cycle progression in glioma cells via extracellular signal-regulated kinase activation. Cancer Res 2000; 60:1766-1772.
66. Desbarats J, Birge RB, Mimouni-Rongy M et al. Fas engagement induces neurite growth through ERK activation and p35 upregulation. Nat Cell Biol 2003; 5:118-125.
67. Lambert C, Landau AM, Desbarats J. Fas-beyond death: A regenerative role for Fas in the nervous system. Apoptosis 2003; 8:551-562.
68. Ruckenstein MJ, Milburn M, Hu L. Strial dysfunction in the MRL-Fas mouse. Otolaryngol Head Neck Surg 1999; 121:452-456.
69. Ruckenstein MJ, Sarwar A, Hu L et al. Marion TN. Effects of immunosuppression on the development of cochlear disease in the MRL-Fas(lpr) mouse. Laryngoscope 1999; 109:626-630.
70. Hess DC, Taormina M, Thompson J et al. Cognitive and neurologic deficits in the MRL/lpr mouse: A clinicopathologic study. J Rheumatol 1993; 20:610-617.
71. Moalem G, Leibowitz-Amit R, Yoles E et al. Autoimmune T cells protect neurons from secondary degeneration after central nervous system axotomy. Nat Med 1999; 5:49-55.
72. Lucas R, Holmgren L, Garcia I et al. Multiple forms of angiostatin induce apoptosis in endothelial cells. Blood 1998; 92:4730-4741.
73. Jimenez B, Volpert OV, Crawford SE et al. Signals leading to apoptosis-dependent inhibition of neovascularization by thrombospondin-1. Nat Med 2000; 6:41-48.
74. Stellmach V, Crawford SE, Zhou W et al. Prevention of ischemia-induced retinopathy by the natural ocular antiangiogenic agent pigment epithelium-derived factor. Proc Natl Acad Sci USA 2001; 98:2593-2597.
75. Volpert OV, Zaichuk T, Zhou W et al. Inducer-stimulated Fas targets activated endothelium for destruction by anti-angiogenic thrombospondin-1 and pigment epithelium-derived factor. Nat Med 2002; 8:349-357.
76. Zaichuk TA, Shroff EH, Emmanuel R et al. Nuclear factor of activated T cells balances angiogenesis activation and inhibition. J Exp Med 2004; 199:1513-1522.
77. Kamphaus GD, Colorado PC, Panka DJ et al. Canstatin, a novel matrix-derived inhibitor of angiogenesis and tumor growth. J Biol Chem 2000; 275:1209-1215.
78. Panka DJ, Mier JW. Canstatin inhibits Akt activation and induces Fas-dependent apoptosis in endothelial cells. J Biol Chem 2003; 278:37632-37636.
79. Filippatos G, Leche C, Sunga R et al. Expression of FAS adjacent to fibrotic foci in the failing human heart is not associated with increased apoptosis. Am J Physiol 1999; 277:H445-451.
80. Badorff C, Ruetten H, Mueller S et al. Fas receptor signaling inhibits glycogen synthase kinase 3 beta and induces cardiac hypertrophy following pressure overload. J Clin Invest 2002; 109:373-381.
81. Choukroun G, Hajjar R, Fry S et al. Regulation of cardiac hypertrophy in vivo by the stress-activated protein kinases/c-Jun NH(2)-terminal kinases. J Clin Invest 1999; 104:391-398.
82. Esposito G, Prasad SV, Rapacciuolo A et al. Cardiac overexpression of a G(q) inhibitor blocks induction of extracellular signal-regulated kinase and c-Jun NH(2)-terminal kinase activity in vivo pressure overload. Circulation 2001; 103:1453-1458.
83. Watanabe-Fukunaga R, Brannan CI, Copeland NG et al. Lymphoproliferation disorder in mice explained by defects in Fas antigen that mediates apoptosis. Nature 1992; 356:314-317.
84. Takahashi T, Tanaka M, Brannan CI et al. Generalized lymphoproliferative disease in mice, caused by a point mutation in the Fas ligand. Cell 1994; 76:969-976.
85. Rieux-Laucat F, Fischer A, Deist FL. Cell-death signaling and human disease. Curr Opin Immunol 2003; 15:325-331.
86. Siegel RM, Chan FK, Chun HJ et al. The multifaceted role of Fas signaling in immune cell homeostasis and autoimmunity. Nat Immunol 2000; 1:469-474.

87. Sabelko KA, Kelly KA, Nahm MH et al. Fas and Fas ligand enhance the pathogenesis of experimental allergic encephalomyelitis, but are not essential for immune privilege in the central nervous system. J Immunol 1997; 159:3096-3099.

88. Waldner H, Sobel RA, Howard E et al. Fas- and FasL-deficient mice are resistant to induction of autoimmune encephalomyelitis. J Immunol 1997; 159:3100-3103.

89. Malipiero U, Frei K, Spanaus KS et al. Myelin oligodendrocyte glycoprotein-induced autoimmune encephalomyelitis is chronic/relapsing in perforin knockout mice, but monophasic in Fas- and Fas ligand-deficient lpr and gld mice. Eur J Immunol 1997; 27:3151-3160.

90. Sabelko-Downes KA, Cross AH, Russell JH. Dual role for Fas ligand in the initiation of and recovery from experimental allergic encephalomyelitis. J Exp Med 1999; 189:1195-1205.

91. Dittel BN, Merchant RM, Janeway Jr CA. Evidence for Fas-dependent and Fas-independent mechanisms in the pathogenesis of experimental autoimmune encephalomyelitis. J Immunol 1999; 162:6392-6400.

92. Okuda Y, Sakoda S, Fujimura H et al. Intrathecal administration of neutralizing antibody against Fas ligand suppresses the progression of experimental autoimmune encephalomyelitis. Biochem Biophys Res Commun 2000; 275:164-168.

93. Wildbaum G, Westermann J, Maor G et al. A targeted DNA vaccine encoding fas ligand defines its dual role in the regulation of experimental autoimmune encephalomyelitis. J Clin Invest 2000; 106:671-679.

94. Suvannavejh GC, Dal Canto MC, Matis LA et al. Fas-mediated apoptosis in clinical remissions of relapsing experimental autoimmune encephalomyelitis. J Clin Invest 2000; 105:223-231.

95. Galle PR, Hofmann WJ, Walczak H et al. Involvement of the CD95 (APO-1/Fas) receptor and ligand in liver damage. J Exp Med 1995; 182:1223-1230.

96. Kondo T, Suda T, Fukuyama H et al. Essential roles of the Fas ligand in the development of hepatitis. Nat Med 1997; 3:409-413.

97. Chisari FV. Cytotoxic T cells and viral hepatitis. J Clin Invest 1997; 99:1472-1477.

98. Tsai SL, Huang SN. T cell mechanisms in the immunopathogenesis of viral hepatitis B and C. J Gastroenterol Hepatol 1997; 12:227-235.

99. Hayashi N, Mita E. Fas system and apoptosis in viral hepatitis. J Gastroenterol Hepatol 1997; 12:223-226.

100. Mochizuki K, Hayashi N, Hiramatsu N et al. Fas antigen expression in liver tissues of patients with chronic hepatitis B. J Hepatol 1996; 6:1-7.

101. Yoo YG, Lee JS, Lee MO. Hepatitis B virus X protein induces expression of fas ligand gene through enhancing transcriptional activity of early growth response factor. J Biol Chem 2004, Epub ahead of print.

102. Lee SH, Shin MS, Lee HS et al. Expression of Fas and Fas-related molecules in human hepatocellular carcinoma. Hum Pathol 2001; 32(3):250-256.

103. Tsutsui H, Kayagaki N, Kuida K et al. Caspase-1-independent, Fas/Fas ligand-mediated IL-18 secretion from macrophages causes acute liver injury in mice. Immunity 1999; 11:359-367.

104. Seino K, Kayagaki N, Takeda K et al. Contribution of Fas ligand to T cell-mediated hepatic injury in mice. Gastroenterology 1997; 113:1315-1322.

105. Tagawa Y, Kakuta S, Iwakura Y. Involvement of Fas/Fas ligand system-mediated apoptosis in the development of concanavalin A-induced hepatitis. Eur J Immunol 1998; 28:4105-4113.

106. Adachi M, Suematsu S, Kondo T et al. Targeted mutation in the Fas gene causes hyperplasia in peripheral lymphoid organs and liver. Nat Genet 1995; 11:294-300.

107. Desbarats J, Newell MK. Fas engagement accelerates liver regeneration after partial hepatectomy. Nat Med 2000; 6:920-923.

108. Raoul C, Henderson CE, Pettmann B. Programmed cell death of embryonic motoneurons triggered through the Fas death receptor. J Cell Biol 1999; 147:1049-1062.

109. Ugolini G, Raoul C, Ferri A et al. Fas/tumor necrosis factor receptor death signaling is required for axotomy-induced death of motoneurons in vivo. J Neurosci 2003; 23:8526-8531.

110. Raoul C, Estevez AG, Nishimune H et al. Motoneuron death triggered by a specific pathway downstream of Fas. potentiation by ALS-linked SOD1 mutations. Neuron 2002; 35:1067-1083.

111. Demjen D, Klussmann S, Kleber S et al. Neutralization of CD95 ligand promotes regeneration and functional recovery after spinal cord injury. Nat Med 2004; 10:389-395.

112. Holler N, Kataoka T, Bodmer JL et al. Development of improved soluble inhibitors of FasL and CD40L based on oligomerized receptors. J Immunol Methods 2000; 237:159-173.

113. Hasegawa A, Cheng X, Kajino K et al. Fas-disabling small exocyclic peptide mimetics limit apoptosis by an unexpected mechanism. Proc Natl Acad Sci USA 2004; 101:6599-6604.

114. Ogasawara J, Watanabe-Fukunaga R, Adachi M et al. Lethal effect of the anti-Fas antibody in mice. Nature 1993; 364:806-809.

115. Nishimura Y, Hirabayashi Y, Matsuzaki Y et al. In vivo analysis of Fas antigen-mediated apoptosis: Effects of agonistic anti-mouse Fas mAb on thymus, spleen and liver. Int Immunol 1997; 9:307-316.

116. Nishimura-Morita Y, Nose M, Inoue T et al. Amelioration of systemic autoimmune disease by the stimulation of apoptosis-promoting receptor Fas with anti-Fas mAb. Int Immunol 1997; 9:1793-1799.

117. Barnhart BC, Alappat EC, Peter ME. The CD95 type I/type II model. Semin Immunol 2003; 15:185-193.

118. Ichikawa K, Yoshida-Kato H, Ohtsuki M et al. A novel murine anti-human Fas mAb which mitigates lymphadenopathy without hepatotoxicity. Int Immunol 2000; 12:555-562.

119. Lau HT, Yu M, Fontana A et al. Prevention of islet allograft rejection with engineered myoblasts expressing FasL in mice. Science 1996; 273:109-112.

120. Allison J, Georgiou HM, Strasser A et al. Transgenic expression of CD95 ligand on islet beta cells induces a granulocytic infiltration but does not confer immune privilege upon islet allografts. Proc Natl Acad Sci USA 1997; 94:3943-3947.

121. Takeda Y, Gotoh M, Dono K et al. Protection of islet allografts transplanted together with Fas ligand expressing testicular allografts. Diabetologia 1998; 41:315-321.

122. Li XC, Li Y, Dodge I et al. Induction of allograft tolerance in the absence of Fas-mediated apoptosis. J Immunol 1999; 163:2500-2507.

123. Turvey SE, Gonzalez-Nicolini V, Kingsley CI et al. Fas ligand-transfected myoblasts and islet cell transplantation. Transplantation 2000; 69:1972-1976.

124. Zhang H, Yang Y, Horton JL et al. Amelioration of collagen-induced arthritis by CD95 (Apo-1/Fas)-ligand gene transfer. J Clin Invest 1997; 100:1951-1957.

125. Kim SH, Kim S, Oligino TJ et al. Effective treatment of established mouse collagen-induced arthritis by systemic administration of dendritic cells genetically modified to express FasL. Mol Ther 2002; 6:584-590.

126. Guery L, Batteux F, Bessis N et al. Expression of Fas ligand improves the effect of IL-4 in collagen-induced arthritis. Eur J Immunol 2000; 30:308-315.

127. Morelli AE, Larregina AT, Smith-Arica J et al. Neuronal and glial cell type-specific promoters within adenovirus recombinants restrict the expression of the apoptosis-inducing molecule Fas ligand to predetermined brain cell types, and abolish peripheral liver toxicity. J Gen Virol 1999; 80(Pt 3):571-583.

128. Ambar BB, Frei K, Malipiero U et al. Treatment of experimental glioma by administration of adenoviral vectors expressing Fas ligand. Hum Gene Ther 1999; 10:1641-1648.

129. Aoki K, Akyurek LM, San H et al. Restricted expression of an adenoviral vector encoding Fas ligand (CD95L) enhances safety for cancer gene therapy. Mol Ther 2000; 1:555-565.

130. Hyer ML, Voelkel-Johnson C, Rubinchik S et al. Intracellular Fas ligand expression causes Fas-mediated apoptosis in human prostate cancer cells resistant to monoclonal antibody-induced apoptosis. Mol Ther 2000; 2:348-358.

131. Rubinchik S, Ding R, Qiu AJ et al. Adenoviral vector which delivers FasL-GFP fusion protein regulated by the tet-inducible expression system. Gene Ther 2000; 7:875-885.

132. Rubinchik S, Wang D, Yu H et al. A complex adenovirus vector that delivers FASL-GFP with combined prostate-specific and tetracycline-regulated expression. Mol Ther 2001; 4:416-426.

133. Friesen C, Herr I, Krammer PH et al. Involvement of the CD95 (APO-1/FAS) receptor/ligand system in drug-induced apoptosis in leukemia cells. Nat Med 1996; 2:574-577.

134. Fulda S, Sieverts H, Friesen C et al. The CD95 (APO-1/Fas) system mediates drug-induced apoptosis in neuroblastoma cells. Cancer Res 1997; 57:3823-3829.

135. Muller M, Strand S, Hug H et al. Drug-induced apoptosis in hepatoma cells is mediated by the CD95 (APO-1/Fas) receptor/ligand system and involves activation of wild-type p53. J Clin Invest 1997; 99:403-413.

136. Trauth BC, Klas C, Peters AM et al. Monoclonal antibody-mediated tumor regression by induction of apoptosis. Science 1989; 245:301-305.

137. Shimizu M, Yoshimoto T, Nagata S et al. A trial to kill tumor cells through Fas (CD95)-mediated apoptosis in vivo. Biochem Biophys Res Commun 1996; 228:375-379.

138. Schneider P, Holler N, Bodmer JL et al. Conversion of membrane-bound Fas(CD95) ligand to its soluble form is associated with downregulation of its proapoptotic activity and loss of liver toxicity. J Exp Med 1998; 187:1205-1213.

139. Samel D, Muller D, Gerspach J et al. Generation of a FasL-based proapoptotic fusion protein devoid of systemic toxicity due to cell-surface antigen-restricted Activation. J Biol Chem 2003; 278:32077-32082.

140. Wajant H, Moosmayer D, Wuest T et al. Differential activation of TRAIL-R1 and -2 by soluble and membrane TRAIL allows selective surface antigen-directed activation of TRAIL-R2 by a soluble TRAIL derivative. Oncogene 2001; 20:4101-4106.

141. Wuest T, Gerlach E, Banerjee D et al. TNF-Selectokine: A novel prodrug generated for tumor targeting and site-specific activation of tumor necrosis factor. Oncogene 2002; 21:4257-4265.

142. Jung G, Grosse-Hovest L, Krammer PH et al. Target cell-restricted triggering of the CD95 (APO-1/ Fas) death receptor with bispecific antibody fragments. Cancer Res 2001; 61:1846-1848.

143. Hagen TL, Eggermont AM. Tumor vascular therapy with TNF: Critical review on animal models. Methods Mol Med 2004; 98:227-246.

144. Grunhagen DJ, Brunstein F, ten Hagen TL et al. TNF-based isolated limb perfusion: A decade of experience with antivascular therapy in the management of locally advanced extremity soft tissue sarcomas. Cancer Treat Res 2004; 120:65-79.

CHAPTER 11

Tools for Activation and Neutralization of Fas Signaling

Pascal Schneider

Abstract

Apoptosis mediated by the Fas / FasL pair of receptor and ligand is involved in physiological or pathological processes in which cell death is either required to eliminate potentially harmful infected, transformed or autoreactive cells, or is detrimental when healthy cells like hepatocytes are massively killed. Since the discovery of cytotoxic anti-Fas antibodies, agonists and antagonists of Fas and FasL have not only proven essential tools for this field of research, but also provided clues on the molecular mechanism of Fas activation. In order to signal apoptosis, Fas not only requires to be engaged by FasL, but also needs to be so in a multivalent fashion. Hence, agents that prevent higher order multimerization of Fas act as antagonists, even if they do bind Fas. The inherent resistance of a cell to suboptimal Fas signals is another essential factor regarding the outcome of Fas engagement. This review discusses agonists and antagonists of the Fas pathway that act at the level of the extracellular domain of either Fas or FasL.

Introduction

Apoptosis mediated by Fas was discovered more than a decade ago by the development of two antibodies able to rapidly induce cell death in a variety of human cell lines.[1,2] These antibodies were raised independently against plasma membrane component of the SKW6.4 B human lymphoblastic cell line in one instance, and against the FS-7 fibroblasts in the other instance, and were called anti-Apo-1 and anti-Fas (CH11), respectively. One of these antibodies, anti-Apo-1, induced spectacular regression of human tumor xenografts in the mouse, suggesting it could be used as an anti-cancer therapeutic agent.[1] The antigen recognized by these antibodies, Fas (also known as Apo-1, CD95 and TNFRSF6), and its natural ligand, FasL, were subsequently cloned in human and mouse, and recognized as members of the TNF and TNF receptor families.[3-6] The development of an anti-mouse Fas agonistic antibody (Jo2) dampened the hope of using Fas as an anti-tumor target. Indeed, intraperitoneal administration of Jo2 was lethal to mice within a few hours because of the exquisite sensitivity of normal hepatocytes to anti-Fas antibodies.[7]

The role of Fas and FasL in the effector function and homeostasis of the immune system was established upon identification of genetic defects affecting Fas or FasL in mouse and human and resulting in lymphoproliferation and autoimmunity (reviewed in refs. 8,9). FasL is also involved in the maintenance of immonoprivileged sites and plays a proinflammatory roles

Fas Signaling, edited by Harald Wajant. ©2006 Landes Bioscience and Springer Science+Business Media.

in certain circumstances.[10] Fas transduces the apoptotic signal through the formation of an activation platform on which pro-caspases 8 and 10 are processed to their mature, pro-apoptotic forms of intracellular cysteine proteases (reviewed in ref. 11).

The ability to manipulate the Fas system both in vitro and in vivo has proven important for many studies. When looking at the reagents available to positively or negatively interfere with Fas-induced apoptosis, a recurrent theme emerges, which is the extent to which the reagent can or cannot cross-link Fas above a certain threshold required for the induction of apoptosis, the level of this threshold being a characteristic of a given cell at a given time. FasL assembles as homotrimers with three receptor binding sites and can be found in two main forms: "soluble FasL", which consists only of trimeric entities, and "membrane-bound FasL" or "membrane-bond-like FasL", in which more than one FasL trimers are somehow held together. For most cells, soluble FasL does not cross-link Fas sufficiently to induce apoptosis, whereas membrane-bound-like FasL does. However, nontoxic soluble FasL becomes pro-apoptotic if it is converted to a membrane-bound-like FasL by cross-linking. Thus, if multiple Fas/FasL complexes are formed in close proximity, the pro-caspase-8 units recruited to the receptor can cross-activate themselves and initiate apoptosis.

The present review discusses in a nonexhaustive manner how to experimentally induce Fas-dependent apoptosis or to block the pathway with agents acting on Fas or FasL (Fig. 1). Information on Fas and FasL reagents not cited in this review can be found on various sites, e.g., http://staging.alzforum.org/res/com/ant/FAS/FAS-tableC.html#kamMC073 or http://www.vincibiochem.it/FasFasL.htm .

Fas Agonists

Agonistic Anti-Human Fas Antibodies

Anti-Apo-1 is a mouse monoclonal IgG3 directed against human Fas. This antibody has been widely used in vitro and in vivo owing to its ability to induce apoptosis in Fas positive cells.[1] In contrast to the IgG3, isotype switch variants of this antibody display little (IgG1 and IgA) or no proapoptotic activity (IgG2a and IgG2b), unless they are cross-linked by an other antibody.[12] This suggests that aggregation of the antibody, either through Fc - Fc interactions in the IgG3 isotype or via secondary antibodies, is necessary to induce Fas signaling. This notion is reinforced by the analysis of a pair of antibodies that recognize the same epitope of human Fas (amino acids 116-135 EINCTRQNTKCRCKPNFFC).[13] The pentameric CH11 IgM antibody is proapoptotic,[2] but the ZB4 IgG1 monoclonal antibody or a divalent Fab'2 fragment of CH11 are not. Although the oligomerization of an anti-Fas antibody appears beneficial for induction of apoptosis, other parameters such as the relative rigidity of the hinge region of IgG3, or differences in affinity or epitopes have been proposed to modulate the agonistic properties of anti-Fas antibodies.[12,14]

An adequate conformation and/or preassociation of Fas may also be of importance.[15] A point mutation (H60R) in Fas confers complete resistance to CH11-mediated apoptosis, without affecting the epitope recognized by the antibody.[16] It is conceivable that this mutation disrupts the preligand association domain of Fas, making it less responsive to activation by cross-linking.

Other human Fas agonists include the mouse IgG2a SM1/1 (that requires cross-linking)[17] (and antibody datasheet) and the mouse IgG3 2R2 (whose toxicity is further enhanced by cross-linking)[18,19] (and antibody datasheet).

Of note, agonistic anti-Fas autoantibodies can form naturally in the course of HIV infection. This is probably due to the fact that an epitope of HIV gp120 corresponds precisely to the amino acid sequence 115-121 of Fas (VEINTCTR).[20] Interestingly, this sequence is part of the CH11 epitope.[14] These autoantibodies have been proposed to contribute to the depletion of CD4+ T cells.

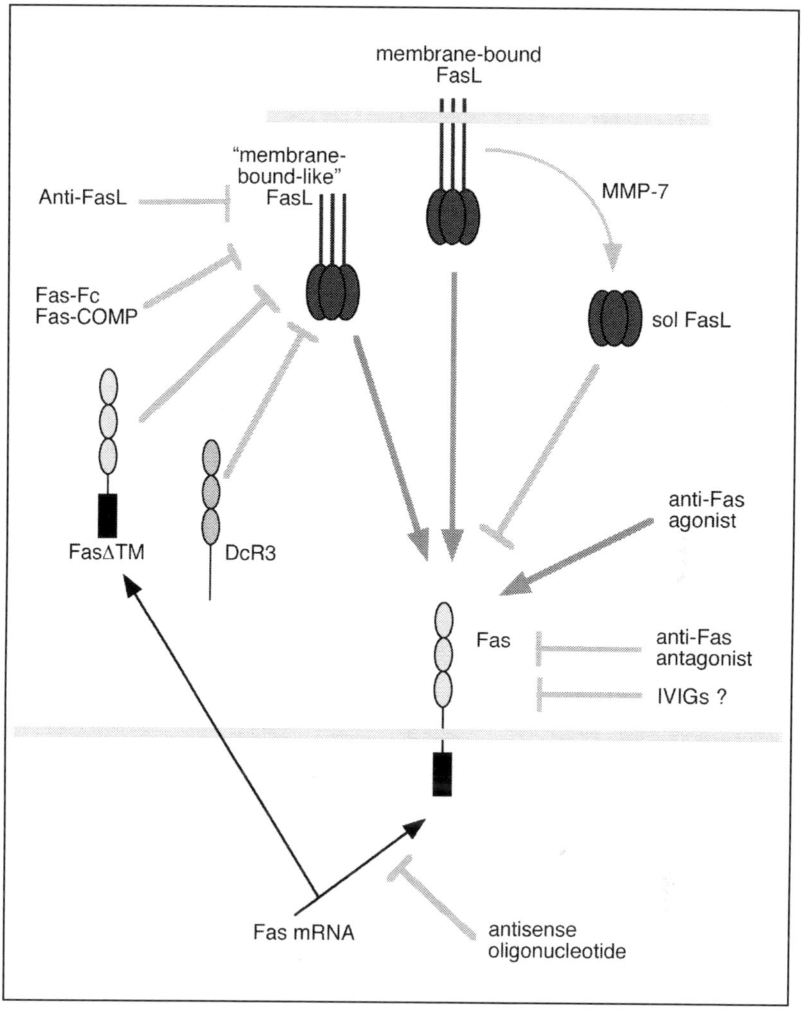

Figure 1. Agonists and antagonists of the Fas / FasL system. FasL exists in two main forms. One is active (membrane-bound FasL and its analogues, "membrane-bound-like" FasL) and the other has little if any cytotoxic activity ("soluble" FasL). Matrixmetalloprotease-7 (MMP-7) is one of the metalloproteases mediating the conversion of membrane-bound FasL to its soluble form. FasL antagonists consist mainly of neutralizing antibodies and soluble receptors. Fas-Fc and Fas-COMP are engineered protein containing the extracellular domain of Fas fused to an oligomerizing domain. Apoptosis is triggered by membrane-bound and "membrane-bound-like" FasL and by some antibodies. Reagents that bind Fas without inducing extensive cross-linking act as antagonists (soluble FasL, some antibodies). Intravenous immunoglobulins (IVIGs) may contain antagonistic anti-Fas antibodies. The antisense Fas oligonucleotide lowers the level of Fas mRNA.

Agonistic Anti-Mouse Fas Antibodies

The Jo2 antibody is an hamster IgG directed against the extracellular domain of mouse Fas, which is famous for its astonishing ability to kill mice within a few hours after intraperitoneal injection.[7] Death is not the direct result of hepatocyte apoptosis, but of collateral damage to blood vessels that lead to liver hemorrhage.

RK-8 is another hamster antibody targeted against mouse Fas that induces apoptosis in Fas positive target cells. However, Jo2 and RK-8 are different in several respects. Whereas Jo2 reacts with all mouse Fas, RK-8 recognizes Fas from Balb/c and MRL strains, but only weakly that of C3H strain.[21] This correlates with a mouse Fas polymorphism in which *mfas.1* (C3H, B6 DBA/1 and DBA/2) has an Arg and *mfas.2* (Balb/c and MRL) a His at position 17. In addition, the epitope of RK-8 is outside the FasL binding site, whereas Jo2 competes for binding with FasL.[22] In vitro, Jo2 is more toxic towards primary hepatocytes and RK-8 is more potent at inducing cell death in lymphoid cells. Although high doses of RK-8 mediate as much or even more liver damage as Jo2 (as measured by the release of liver transaminases in the circulation) it does not induce hemorrhage in adult animals and is not lethal.[22] The RK-8 antibody has been used in vivo to deplete thymocytes or other Fas-sensitive cells such as autoreactive T cells mediating collagen type-II-induced arthritis.[22,23]

Recombinant Fas Ligand

The cytolytic activity of recombinant FasL has been a confusing field, and it is hard to believe that titles as different as "the functional soluble form of FasL" and "soluble FasL blocks the killing" can have been published by the same group using the exact same FasL.[24,25] This is due to the fact that the cytolytic activity of FasL depends on the nature of both the ligand and the target cell. Two main factors are thus involved to decide whether or not Fas triggering results in apoptosis. The first one is the "aggregation" state of FasL trimers: a trimeric soluble FasL comprising the TNF homology domain only is three order of magnitude less active on Jurkat target cells than the same FasL cross-linked by an anti-tag antibody.[26] Thus, conversion of a "soluble" FasL to a "membrane-bound-like" FasL transforms a weak proapoptotic signal into a strong one, both in vitro and in vivo. The second parameter is the intrinsic ability of a target cell to cope with weak Fas signals, which may be related to the ratio between Fas and intracellular inhibitors of the pathway such as FLIP.[27] Thus, suboptimal signals induced by a trimeric soluble human FasL is sufficient to kill the mouse T lymphoma WR19L transfected with murine Fas, and human ConA blasts, but is inactive on Jurkat cells, peripheral blood T cells or mouse hepatocytes. In contrast, membrane-bound FasL kills them all.[24,25,28]

As mentioned above, Flag-tagged soluble FasL is rendered active by cross-linking with an anti-Flag antibody (sometimes referred to as "enhancer").[26] This effect does not take place at high FasL to antibody ratio, probably because the divalent antibody preferentially engages two Flag epitopes within the same FasL trimer.[29] Cross-linking only occurs when a second antibody binds to the third Flag epitope and recruits a second trimer into the complex (Fig. 2). It is also possible to bring two trimeric FasL within the same molecule by fusion with oligomerization domains such as the collagen domain of ACRP30/adiponectin (mega-ligand) or the Fc domain of human IgG1 (Fc-ligand). These molecules have excellent specific activities on a variety of cell lines and can be considered good mimics of membrane-bound FasL.[29]

If two trimers of FasL must act in close proximity to induce apoptosis, how is this achieved when FasL is membrane-bound? The membrane augments the local concentration of FasL by restricting it within a plan and normalizes its spatial orientation. However, the stalk of FasL (i.e., the portion of the ligand located between the transmembrane segment and the TNF homology domain) may have some self-aggregating properties, which would help maintain clusters of FasL within the membrane. Soluble murine FasL containing the stalk region has an apparent molecular weight consistent with an hexameric structure and is far more active than its trimeric counterpart containing the TNF homology domain only,[24,30] supporting the notion that the stalk region may mediate interaction between trimers.

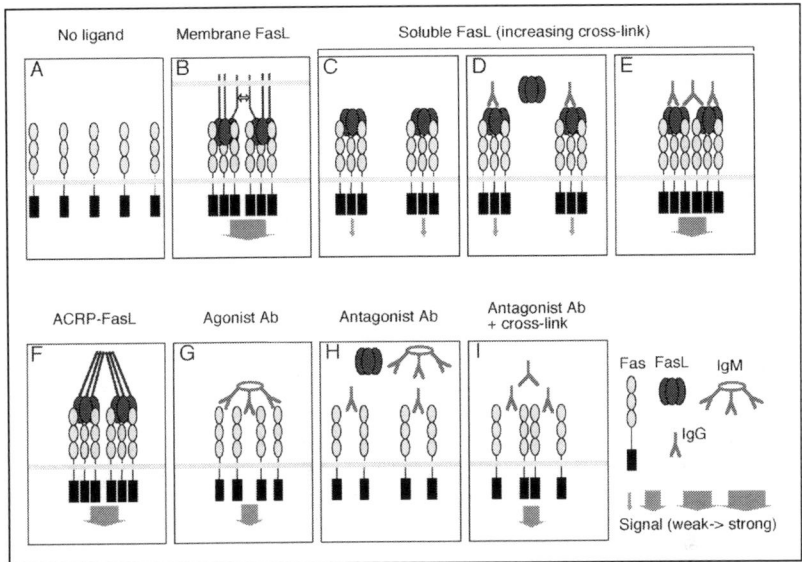

Figure 2. Model of Fas engagement by agonists and antagonists. A) In the absence of ligand, Fas does not deliver an apoptotic signal in the cell. B) Membrane-bound FasL is an adequate agonist of Fas. Stalk-stalk interactions may favor clusters of FasL in the membrane. C) Soluble, trimeric FasL induces a weak signal. Most cells have a higher threshold of activation and do not undergo apoptosis. Thus, for most cells, soluble FasL acts as an antagonist. D) If the ratio of soluble FasL to cross-linking (anti-tag) antibody is high, the antibody binds two tags within the same trimer of FasL. This does not induce cross-linking of trimers and cells are protected. E) If the ratio of soluble FasL to cross-linking antibody is low, FasL trimers are cross-linked and become potent Fas agonists. F) Recombinant proteins engineered to contain two trimers of FasL (ACRP-FasL, Fc-FasL) mimic membrane-bound FasL and are proapoptotic. G) Agonistic anti-FasL antibodies cross-link sufficient Fas molecules to induce a signal. This signal is however not as optimal as that of membrane-bound FasL, and fails to kill cells with high intrinsic threshold to Fas signaling. H) Antagonist antibodies do not cross-link Fas molecules sufficiently enough to kill the cell but make binding sites unavailable for other signaling competent molecules. In order not to be displaced by agonists, antagonist antibodies must be of high affinity. I) An antagonist antibody that is cross-linked (by an anti-antibody, by self-association or by any other aggregation mechanism) becomes an agonist.

Microvesicles and Particles

Cells expressing surface FasL are potent Fas agonists. However, lymphoid cells direct FasL to secretory lysozomes via retention signals comprising the proline-rich region of its intracellular domain.[31] Significant constitutive surface expression of FasL is only achieved by deleting this region or by transfecting FasL in other cell types.[32] Supernatants of FasL transfected cells can contain proteolytically processed FasL,[33] but have also been reported to contain full length and active FasL associated with membrane vesicles, although the latter were poorly characterized.[26,34] Better defined particles containing membrane-bound FasL have been generated that are potent inducers of cell death.[35] In this system, FasL transfected NIH 3T3 cells were used to package Moloney murine leukemia virus (MMLV). FasL expressed by NIH cells gets incorporated into the lipid bilayer of the retrovirus as it buds out of the packaging cell, generating FasL-coated particles that are cytotoxic to Fas expressing cells. Favoring the contact between the retroviral particle and the target cell with cationic compounds such as Polybrene or poly-L-lysine enhanced the killing.[35] Use of these FasL-containing vesicles will not only kill Fas sensitive cells, but also transfer a FasL gene into the target cell, which may not always be desirable.

Target-Induced Multimerization of FasL

The observation that soluble FasL is not or only poorly toxic, but that cross-linked FasL is toxic suggests that specific induction of soluble FasL aggregation on a target cell would allow to kill this cell without affecting its neighbors. This is achieved by fusing soluble, trimeric, inactive FasL to a receptor or to a single chain antibody. When the receptor or the single chain antibody recognizes its cognate ligand, FasL shifts from its soluble to the membrane-bound-like form, allowing specific killing of the target cell. With these reagents, antigen-positive tumor cells can be killed in a live animal without inducing the liver toxicity inherent to other Fas agonists.[36]

Inhibitors of the Fas-FasL System

Soluble Fas

A common strategy with receptors of the TNF family is to couple the extracellular domain of a receptor to the Fc portion of an antibody, thereby generating a dimeric, soluble decoy receptor.[37] Although Fas-Fc binds FasL and can inhibit FasL-mediated cells death, it does so relatively poorly in comparison with blocking anti-FasL antibodies.[32,33,38] Despite this, Fas-Fc remains a useful tool to antagonize FasL both in vitro and in vivo if used at high enough concentrations. The affinity of Fas-Fc for FasL is relatively low compared to other receptor ligand pairs, explaining its suboptimal antagonist activity. The antagonist effect of Fas-Fc was improved by cross-linking with protein-A, or by replacing the Fc portion with the pentamerization domain of cartilage oligomeric matrix protein to yield Fas-COMP, an improved inhibitor with higher avidity for FasL that was as potent or better than neutralizing anti-FasL antibodies.[39]

Both human Fas-Fc and mouse Fas-Fc have been used successfully in vivo to inhibit hepatitis, graft versus host disease, and to protect immunoregulatory T cells from cyclophosphamide toxicity.[38,40-42]

A natural form of soluble Fas (FasΔTM) is generated by alternative splicing of exon 6 encoding the transmembrane domain.[43] This form of soluble Fas is elevated in sera of patients with autoimmune diseases or cancer. One study suggests that FasΔTM can not only oligomerize and inhibit FasL, but also induce a deleterious "retrosignal" in FasL expressing cells, which would be an elegant manner to suppress both FasL and its source.[44]

DcR3

In addition to FasΔTM, FasL has been reported to bind another soluble decoy receptor, coined DcR3 (or TR6).[45] DcR3 also binds the ligands Light and TL1A.[46,47] DcR3 harbors a protease cleavage site in the sequence TPR-AGR, that is cleaved in vitro by thrombin and in vivo upon subcutaneous injection.[48] The cleavage separates the cysteine-rich domain of DcR3 from its C-terminal extension. Remarkably, this cleavage abolishes DcR3 binding to FasL, without affecting its ability to inhibit Light. A non cleavable variant of DcR3 (R247Q) has been used in vivo to block FasL-induced neutrophil infiltration in the lung.[48,49] It will be of interest to understand how the C-terminal portion of DcR3 participates in the binding of FasL.

Soluble FasL

Any reagent that binds to Fas without inducing optimal signaling is susceptible to protect cells against a better Fas agonist. This is the case of trimeric human FasL which protected peripheral blood T cells or HeLa cells against active versions of FasL.[24,29] This reagent is predicted to work better with cells having a higher threshold to Fas-induced apoptosis.

Blocking Anti-Fas Antibodies

Several blocking anti-human Fas antibodies have been described, of which mAb ZB4 (mouse IgG1) is an interesting example. ZB4 is an efficient blocker of apoptosis induced by CH11 IgM.[50] Both antibodies recognize the same (or closely overlapping) epitope located on the outer surface of Fas, i.e., the surface that does not directly interact with FasL.[14] Nevertheless, ZB4 and FasL bind to Fas in a mutually exclusive manner.[29] Antibodies with the same specificity and isotype as ZB4, but with lower affinity do not antagonize CH11, suggesting they might be displaced by the latter antibody.[14,51] Of note, ZB4 is an antagonist only in its uncross-linked form, and induces apoptosis upon cross-linking.

The notion of antagonist anti-Fas antibody depends on the target cell under consideration. Both CH11 and Jo2, which are predominantly known as inducers of apoptosis, can act as antagonists of FasL on peripheral blood T cells.[24,30] This indicates that agonistic antibodies are unable to engage Fas with the exact same geometry as membrane-bound FasL.

Fas Antisense Oligonucleotides

Suppression of Fas expression can be mediated in vivo by antisense phosphorothioate oligonucleotides chemically modified with 2'-O-(2-methoxy)ethyl groups at position 1-5 and 15-20 (Fig. 3). These modifications confer enhanced potency to the antisense oligonucleotide and improved resistance to nucleases. Suppression of Fas expression in the liver was achieved at oligonucleotide concentrations greater than 8 μM, corresponding to a daily dose of 20 mg/kg for 4 consecutive days. Following this treatment that reduced Fas expression by 70%, mice resisted administration of a lethal dose of Jo2.[52] Half-life of the oligonucleotide in vivo was close to two weeks, and sufficient depletion of Fas could be maintained for 10 days after oligonucleotide treatment.

This technique has also been applied successfully to downregulate downstream target of Fas signaling such as Bid.[53]

Antisense oligonucleotides have also been used to downregulate FasL expression locally in skin sections.[54]

Blocking Anti-Human FasL Antibodies

The hamster IgG 4H9 and 4A5 and the mouse monoclonal antibodies NOK-1 (IgG1), NOK-2 (IgG2a) and NOK-3 (IgM) interact with distinct epitopes of FasL but they all strongly inhibit FasL activity.[32,33]

The anti-FasL mAb 5G51 (mouse IgG1) has also been used to block glucocorticoid-induced apoptosis, IL10-induced apoptosis of monocytes or activation-induced cell death in Jurkat cells[19,55] (and antibody data sheet).

Autoantibodies against human FasL can develop in the autoimmune condition systemic lupus erythematosus. These antibodies are directed against amino acids 161-170 of FasL that are involved in the structure and receptor binding of FasL.[56] By inhibiting the activity of FasL, these autoantibodies may contribute to the emergence of autoreactive cells.

Blocking Anti-Mouse FasL Antibodies

Murine FasL exists in two allotypes, muFasL.1 (Thr184 and Glu218) and muFasL.2 (Ala184 and Gly218). Curiously, the muFasL.2 variant found in BALB/c, DBA/1 and DBA/2 mice is more active towards both human and murine Fas than muFasL.2 (found in B6, C3H, MRL, SJL, NOD, NZB, NZW mice).[57] The mouse IgG2b mAb K10 (generated in FasL-deficient *gld* mice) blocks FasL.1 but not FasL.2, and the hamster IgG MFL3 inhibits all murine FasL.[57] The hamster IgG FLIM4 and FLIM58 are other anti-mouse FasL blocking antibodies. FLIM58 recognizes both the wild type and the *gld* mutant form of mouse FasL

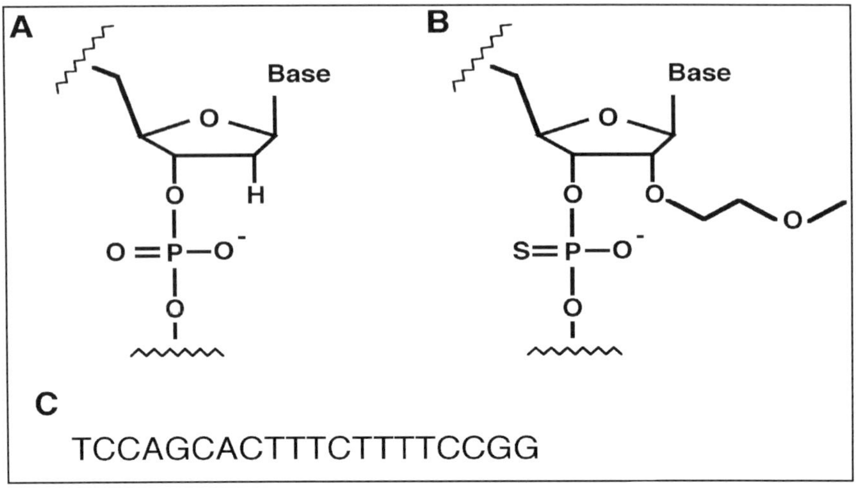

Figure 3. Antisense Fas oligonucleotide. A) Oligonucleotide. B) Oligonucleotide with phosphorothioate and 2'-MOE modifications. These modifications both enhance the effect of the oligonucleotide and reduce its sensitivity to nucleases. C) Sequence of anti-mouse Fas oligonucleotide. This oligonucleotide contains phosphorothioate modifications at all positions, and 2'-MOE modifications at position 1-5 and 15-20.

(Phe273->Leu), whereas FLIM58 recognizes only the wild type. Interestingly, FLIM4 inhibits the activity of soluble recombinant FasL but not that of membrane-bound FasL, whereas FLIM58 blocks both. K10, MFL3 and FLIM58 are effective in vivo. Fas-deficient *lpr* mice express high levels of FasL, which can kill T cells from wild type mice upon transfer. Therefore, *lpr* mice are protected against diabetes induced by the transfer of autoreactive T cells. Neutralization of FasL in *lpr* mice by administration of K10 abolished their resistance to diabetes by preventing killing of the transferred lymphocytes.[58] In another study, the combined use of anti-TNF antibody and MFL3 was effective at reducing brain damage and mortality following induction of stroke in mice.[59] Similarly, FLIM58 inhibited mortality in a model of graft versus host disease.[38]

Matrix Metalloproteinase-7 (MMP-7)

Membrane-bound FasL is released in a soluble form by the action of metalloproteases, one of which is MMP-7. Doxorubicin induced-apoptosis is believed to proceed, at least in part, through upregulation of Fas and FasL in certain tumor cell lines. The effect of doxorubicin was inhibited by Fas-Fc expressed in the target cell itself, or by MMP-7. This suggests that MMP-7 cleaves membrane-bound FasL to a less active form, and could be used as an inhibitor of membrane-bound FasL activity.[60] In vitro assays demonstrated several MMP-7 cleavage sites within the stalk region of FasL, one of which was identified previously in FasL transfected cells.[26,61] The existence of the upstream cleavage sites raises the interesting hypothesis that FasL may be released in longer forms that may retain membrane-bound-like activity, although this remains to be shown.

IVIGs

Toxic epidermal necrolysis (Lyell's syndrome) is a severe acute reaction provoked in certain individuals by normally innocuous drugs. This condition is characterized by the detachment of the epidermis from the dermis and is often fatal despite supportive care on burns or intensive care unit. The drug induces FasL expression in skin, which in turn promote keratinocytes to undergo apoptosis.[62] Treatment of patients with pooled human intravenous immunoglobulins (IVIGs) proved beneficial for both a rapid skin recovery and a favorable outcome in three studies, and death reported in a fourth study were due to the renal failure not directly attributable to the effect of the syndrome.[62-66] It is believed that IVIGs contain a proportion of antagonistic anti-Fas antibodies which may contribute to their therapeutic effect, because IVIGs recognize recombinant Fas by western blot and can protect Fas sensitive cells from apoptosis induced by recombinant FasL.[62,63] It is however unclear to which extent inhibition of Fas / FasL accounts for the therapeutic effect of IVIGs.

Concluding Remarks

The scientific community disposes of valuable and efficient agonists and antagonist of the Fas/FasL system, most of which are commercially available. If the reagents acting on FasL have a relatively straightforward and clear mode of action, the precise molecular mechanism underlying the activity of Fas agonists and antagonists is not entirely understood. This is hardly surprising, given the complexity of Fas signaling for which we do not even know how many Fas molecules are implicated. Fas-binding reagents can provide "null", "weak" or "strong" apoptotic signals depending on how they cross-link Fas, with membrane-bound and membrane-bound-like FasL being the most efficient agonists. Target cells respond to "weak" signals by apoptosis or survival, depending on their intrinsic resistance to Fas signals. Inducers of suboptimal signals can thus kill a target cell and protect another one at the same time. With this in mind, researchers can choose reagents that are best suited to their experimental system and may even take advantage of these differences to specifically reduce undesirable side effects of the reagents.

References

1. Trauth BC, Klas C, Peters AM et al. Monoclonal antibody-mediated tumor regression by induction of apoptosis. Science Jul 21 1989; 245(4915):301-305.
2. Yonehara S, Ishii A, Yonehara M. A cell-killing monoclonal antibody (anti-Fas) to a cell surface antigen codownregulated with the receptor of tumor necrosis factor. J Exp Med May 1 1989; 169(5):1747-1756.
3. Suda T, Takahashi T, Golstein P et al. Molecular cloning and expression of the Fas ligand, a novel member of the tumor necrosis factor family. Cell Dec 17 1993; 75(6):1169-1178.
4. Takahashi T, Tanaka M, Inazawa J et al. Human Fas ligand: Gene structure, chromosomal location and species specificity. Int Immunol Oct 1994; 6(10):1567-1574.
5. Watanabe-Fukunaga R, Brannan CI, Itoh N et al. The cDNA structure, expression, and chromosomal assignment of the mouse Fas antigen. J Immunol Feb 15 1992; 148(4):1274-1279.
6. Itoh N, Yonehara S, Ishii A et al. The polypeptide encoded by the cDNA for human cell surface antigen Fas can mediate apoptosis. Cell Jul 26 1991; 66(2):233-243.
7. Ogasawara J, Watanabe-Fukunaga R, Adachi M et al. Lethal effect of the anti-Fas antibody in mice. Nature Aug 26 1993; 364(6440):806-809.
8. Rieux-Laucat F, Le Deist F, Fischer A. Autoimmune lymphoproliferative syndromes: Genetic defects of apoptosis pathways. Cell Death Differ Jan 2003; 10(1):124-133.
9. Nagata S, Golstein P. The Fas death factor. Science Mar 10 1995; 267(5203):1449-1456.
10. O'Connell J, Houston A, Bennett MW et al. Immune privilege or inflammation? Insights into the Fas ligand enigma. Nat Med Mar 2001; 7(3):271-274.
11. Peter ME, Krammer PH. The CD95(APO-1/Fas) DISC and beyond. Cell Death Differ Jan 2003; 10(1):26-35.
12. Dhein J, Daniel PT, Trauth BC et al. Induction of apoptosis by monoclonal antibody anti-APO-1 class switch variants is dependent on cross-linking of APO-1 cell surface antigens. J Immunol Nov 15 1992; 149(10):3166-3173.
13. Fadeel B, Thorpe J, Chiodi F. Mapping of the linear site on the Fas/APO-1 molecule targeted by the prototypic anti-Fas mAb. Int Immunol Dec 1995; 7(12):1967-1975.

14. Fadeel B, Thorpe CJ, Yonehara S et al. Anti-Fas IgG1 antibodies recognizing the same epitope of Fas/APO-1 mediate different biological effects in vitro. Int Immunol 1997; 9(2):201-209.

15. Siegel RM, Frederiksen JK, Zacharias DA et al. Fas preassociation required for apoptosis signaling and dominant inhibition by pathogenic mutations. Science 2000; 288(5475):2354-2357.

16. Maeda T, Nakayama S, Yamada Y et al. The conformational alteration of the mutated extracellular domain of Fas in an adult T cell leukemia cell line. Biochem Biophys Res Commun Sep 6 2002; 296(5):1251-1256.

17. Gajate C, Mollinedo F. The antitumor ether lipid ET-18-OCH(3) induces apoptosis through translocation and capping of Fas/CD95 into membrane rafts in human leukemic cells. Blood Dec 15 2001; 98(13):3860-3863.

18. Ferrari D, Stepczynska A, Los M et al. Differential regulation and ATP requirement for caspase-8 and caspase-3 activation during CD95- and anticancer drug-induced apoptosis. J Exp Med Sep 7 1998; 188(5):979-984.

19. Schmidt M, Lugering N, Pauels HG et al. IL-10 induces apoptosis in human monocytes involving the CD95 receptor/ligand pathway. Eur J Immunol Jun 2000; 30(6):1769-1777.

20. Silvestris F, Nagata S, Cafforio P et al. Cross-linking of Fas by antibodies to a peculiar domain of gp120 V3 loop can enhance T cell apoptosis in HIV-1-infected patients. J Exp Med Dec 1 1996; 184(6):2287-2300.

21. Nishimura Y, Ishii A, Kobayashi Y et al. Expression and function of mouse Fas antigen on immature and mature T cells. J Immunol May 1 1995; 154(9):4395-4403.

22. Nishimura Y, Hirabayashi Y, Matsuzaki Y et al. In vivo analysis of Fas antigen-mediated apoptosis: effects of agonistic anti-mouse Fas mAb on thymus, spleen and liver. Int Immunol Feb 1997; 9(2):307-316.

23. Ogawa Y, Kuwahara H, Kimura T et al. Therapeutic effect of anti-Fas antibody on a collagen induced arthritis model. J Rheumatol May 2001; 28(5):950-955.

24. Suda T, Hashimoto H, Tanaka M et al. Membrane Fas ligand kills human peripheral blood T lymphocytes, and soluble Fas ligand blocks the killing. J Exp Med 1997; 186(12):2045-2050.

25. Tanaka M, Suda T, Takahashi T et al. Expression of the functional soluble form of human fas ligand in activated lymphocytes. Embo J Mar 15 1995; 14(6):1129-1135.

26. Schneider P, Holler N, Bodmer JL et al. Conversion of membrane-bound Fas(CD95) ligand to its soluble form is associated with downregulation of its proapoptotic activity and loss of liver toxicity. J Exp Med 1998; 187(8):1205-1213.

27. Kataoka T, Ito M, Budd RC et al. Expression level of c-FLIP versus Fas determines susceptibility to Fas ligand-induced cell death in murine thymoma EL-4 cells. Exp Cell Res Feb 15 2002; 273(2):256-264.

28. Tanaka M, Itai T, Adachi M et al. Downregulation of Fas ligand by shedding. Nat Med 1998; 4(1):31-36.

29. Holler N, Tardivel A, Kovacsovics-Bankowski M et al. Two adjacent trimeric Fas ligands are required for Fas signaling and formation of a death-inducing signaling complex. Mol Cell Biol Feb 2003; 23(4):1428-1440.

30. Suda T, Tanaka M, Miwa K et al. Apoptosis of mouse naive T cells induced by recombinant soluble Fas ligand and activation-induced resistance to Fas ligand. J Immunol 1996; 157(9):3918-3924.

31. Blott EJ, Bossi G, Clark R et al. Fas ligand is targeted to secretory lysosomes via a proline-rich domain in its cytoplasmic tail. J Cell Sci Jul 2001; 114(Pt 13):2405-2416.

32. Tanaka M, Suda T, Haze K et al. Fas ligand in human serum. Nature Medicine 1996; 2(3):317-322.

33. Kayagaki N, Kawasaki A, Ebata T et al. Metalloproteinase-mediated release of human Fas ligand. J Exp Med Dec 1 1995; 182(6):1777-1783.

34. Martinez-Lorenzo MJ, Anel A, Gamen S et al. Activated human T cells release bioactive Fas ligand and APO2 ligand in microvesicles. J Immunol Aug 1 1999; 163(3):1274-1281.

35. Jodo S, Strehlow D, Ju ST. Bioactivities of Fas ligand-expressing retroviral particles. J Immunol May 15 2000; 164(10):5062-5069.

36. Samel D, Muller D, Gerspach J et al. Generation of a FasL-based proapoptotic fusion protein devoid of systemic toxicity due to cell-surface antigen-restricted Activation. J Biol Chem Aug 22 2003; 278(34):32077-32082.

37. Ashkenazi A, Chamow SM. Immunoadhesins as research tools and therapeutic agents. Curr Opin Immunol Apr 1997; 9(2):195-200.

38. Miwa K, Hashimoto H, Yatomi T et al. Therapeutic effect of an anti-Fas ligand mAb on lethal graft-versus-host disease. Int Immunol Jun 1999; 11(6):925-931.

39. Holler N, Kataoka T, Bodmer JL et al. Development of improved soluble inhibitors of FasL and CD40L based on oligomerized receptors. J Immunol Methods 2000; 237(1-2):159-173.

40. Mahiou J, Walter U, Lepault F et al. In vivo blockade of the Fas-Fas ligand pathway inhibits cyclophosphamide-induced diabetes in NOD mice. J Autoimmun Jun 2001; 16(4):431-440.

41. Kondo T, Suda T, Fukuyama H et al. Essential roles of the Fas ligand in the development of hepatitis. Nat Med 1997; 3(4):409-413.

42. Ueno Y, Ishii M, Yahagi K et al. Fas-mediated cholangiopathy in the murine model of graft versus host disease. Hepatology Apr 2000; 31(4):966-974.

43. Cheng J, Zhou T, Liu C et al. Protection from Fas-mediated apoptosis by a soluble form of the Fas molecule. Science Mar 25 1994; 263(5154):1759-1762.

44. Proussakova OV, Rabaya NA, Moshnikova AB et al. Oligomerization of soluble Fas antigen induces its cytotoxicity. J Biol Chem Sep 19 2003; 278(38):36236-36241.

45. Pitti RM, Marsters SA, Lawrence DA et al. Genomic amplification of a decoy receptor for Fas ligand in lung and colon cancer. Nature Dec 17 1998; 396(6712):699-703.

46. Yu KY, Kwon B, Ni J et al. A newly identified member of tumor necrosis factor receptor superfamily (TR6) suppresses LIGHT-mediated apoptosis. J Biol Chem May 14 1999; 274(20):13733-13736.

47. Migone TS, Zhang J, Luo X et al. TL1A is a TNF-like ligand for DR3 and TR6/DcR3 and functions as a T cell costimulator. Immunity 2002; 16(3):479-492.

48. Wroblewski VJ, McCloud C, Davis K et al. Pharmacokinetics, metabolic stability, and subcutaneous bioavailability of a genetically engineered analog of DcR3, FLINT [DcR3(R218Q)], in cynomolgus monkeys and mice. Drug Metab Dispos Apr 2003; 31(4):502-507.

49. Wortinger MA, Foley JW, Larocque P et al. Fas ligand-induced murine pulmonary inflammation is reduced by a stable decoy receptor 3 analogue. Immunology Oct 2003; 110(2):225-233.

50. Yonehara S, Nishimura Y, Kishil S et al. Involvement of apoptosis antigen Fas in clonal deletion of human thymocytes. Int Immunol Dec 1994; 6(12):1849-1856.

51. Komada Y, Inaba H, Li QS et al. Epitopes and functional responses defined by a panel of anti-Fas (CD95) monoclonal antibodies. Hybridoma Oct 1999; 18(5):391-398.

52. Zhang H, Cook J, Nickel J et al. Reduction of liver Fas expression by an antisense oligonucleotide protects mice from fulminant hepatitis. Nat Biotechnol Aug 2000; 18(8):862-867.

53. Zhang H, Taylor J, Luther D et al. Antisense oligonucleotide inhibition of Bcl-xL and Bid expression in liver regulates responses in a mouse model of Fas-induced fulminant hepatitis. J Pharmacol Exp Ther Oct 2003; 307(1):24-33.

54. Ji J, Wernli M, Buechner S et al. Fas ligand downregulation with antisense oligonucleotides in cells and in cultured tissues of normal skin epidermis and basal cell carcinoma. J Invest Dermatol Jun 2003; 120(6):1094-1099.

55. Schmidt M, Lugering N, Lugering A et al. Role of the CD95/CD95 ligand system in glucocorticoid-induced monocyte apoptosis. J Immunol Jan 15 2001; 166(2):1344-1351.

56. Mihara S, Suzuki N, Takeba Y et al. Combination of molecular mimicry and aberrant autoantigen expression is important for development of anti-Fas ligand autoantibodies in patients with systemic lupus erythematosus. Clin Exp Immunol. Aug 2002; 129(2):359-369.

57. Kayagaki N, Yamaguchi N, Nagao F et al. Polymorphism of murine Fas ligand that affects the biological activity. Proc Natl Acad Sci USA Apr 15 1997; 94(8):3914-3919.

58. Kim S, Kim KA, Hwang DY et al. Inhibition of autoimmune diabetes by Fas ligand: The paradox is solved. J Immunol Mar 15 2000; 164(6):2931-2936.

59. Martin-Villalba A, Hahne M, Kleber S et al. Therapeutic neutralization of CD95-ligand and TNF attenuates brain damage in stroke. Cell Death Differ 2001; 8(7):679-686.

60. Mitsiades N, Yu WH, Poulaki V et al. Matrix metalloproteinase-7-mediated cleavage of Fas ligand protects tumor cells from chemotherapeutic drug cytotoxicity. Cancer Res Jan 15 2001; 61(2):577-581.

61. Vargo-Gogola T, Crawford HC, Fingleton B et al. Identification of novel matrix metalloproteinase-7 (matrilysin) cleavage sites in murine and human Fas ligand. Arch Biochem Biophys. Dec 15 2002; 408(2):155-161.
62. Viard I, Wehrli P, Bullani R et al. Inhibition of toxic epidermal necrolysis by blockade of CD95 with human intravenous immunoglobulin. Science 1998; 282(5388):490-493.
63. Prins C, Kerdel FA, Padilla RS et al. Treatment of toxic epidermal necrolysis with high-dose intravenous immunoglobulins: Multicenter retrospective analysis of 48 consecutive cases. Arch Dermatol Jan 2003; 139(1):26-32.
64. Trent JT, Kirsner RS, Romanelli P et al. Analysis of intravenous immunoglobulin for the treatment of toxic epidermal necrolysis using SCORTEN: The University of Miami Experience Arch Dermatol Jan 2003; 139(1):39-43.
65. Mayorga C, Torres MJ, Corzo JL et al. Improvement of toxic epidermal necrolysis after the early administration of a single high dose of intravenous immunoglobulin. Ann Allergy Asthma Immunol Jul 2003; 91(1):86-91.
66. Bachot N, Revuz J, Roujeau JC. Intravenous immunoglobulin treatment for Stevens-Johnson syndrome and toxic epidermal necrolysis: A prospective noncomparative study showing no benefit on mortality or progression. Arch Dermatol Jan 2003; 139(1):33-36.

Index